MOLLICUTES AND PLANT DISEASES

By

Dr. Shubhrata R. Mishra

Department of Botany
Vikram University
Ujjain (M.P)

D P H

DISCOVERY PUBLISHING HOUSE
NEW DELHI-110002

First Published-2004

ISBN 81-7141-793-0

Published by

DISCOVERY PUBLISHING HOUSE
4831/24, Ansari Road, Prahlad Street,
Darya Ganj, New Delhi-110002 (India)
Phone: 23279245 • Fax: 91-11-23253475
E-mail:dphtemp@indiatimes.com

Printed at: **Tarun Offset Printers. Delhi-53**

Preface

This book contains a wealth of information on Mollicutes. It provides an interdisciplinary coverage of the up-to-date information on Mollicutes such as Mycoplasma, Spiroplasma, Phytoplasma and plant diseases caused by different Mollicutes. This book is intended to serve postgraduate and graduate students. To the students, this book is not merely a general reference; it is equivalent to several textbooks.

This book is not an outcome of author's original research but is a compilation work incorporating the researches of the scientists in India and abroad. The matter has been compiled from standard texts, reviews, research papers and websites. An attempt was made to obtained comprehensive information covering various aspects in Mollicutes. The first chapter deals with introduction, evolution genomic, and phylogeny of Mollicutes, Noncultivable Mollicutes, Plant pathogenic mollicutes and International Organization for Mycoplasmology (IOM). Second chapter provides taxonomy of Mollicute in which Taxa covered by the ICSP Subcommittee on the Taxonomy of Mollicutes - June 2003, list of names of genera Mycoplasma, Ureaplasma, Entomoplasma, Mesoplasma, Spiroplasma, Acholeplasma, Anaeroplasma, Asteroleplasma as well as Candidatus within the genus Mycoplasma and genus Phytoplasma are given. The NCBI taxonomy for nomenclature or classification of Mollicutes is also given. Chapter 3, 4 and 5 are devoted to detailed information such as general characteristics, structure, important genera, metabolism, cell reproduction cycle, pathogenesis, mechanisms of pathogenicity, different diseases, host defenses, epidemiology, ecology, laboratory diagnosis, control and detecting methods of Mycoplasma, Phytoplasma and Spiroplasma respectively. Chapter 5 is related to discuss about some specific plant diseases caused by different mollicutes. Ample references are supplied at the end of the book to assist the reader in pursuing the subject further.

I am deeply indebted to all authors and publishers of texts, reviews and books from where the matter of this book has been compiled. I wish to express my thanks to Professor M.P. Arora for his editorial advice and guidance. My special thanks are due to my friends, colleagues for their support and helpful suggestions. I am specially thankful to my husband Dr. Ravi Mishra who has generously given advice, information and encouragement on the preparation of this book.

Despite my sincere efforts it is possible that some important information might have escaped my attention. Also, students and teachers might feel the need for further details about mollicutes. I shall be obliged if the readers would send me their valuable suggestions in this regard or any other aspects of the book. These would help me to improve the book.

Dr. Shubhrata R. Mishra

CONTENTS

1

INTRODUCTION

The word "Mycoplasma" is commonly used in reference to the class "Mollicutes" Phylum Firmicutes of the Bacteria. The term "Mollicutes" (mollis, soft; cutis, skin) defines a class of cell wall-less prokaryotes. These include phytoplasmas (infect plants and insects), mycoplasmas (infect animals), spiroplasmas (infect plants and insects), acheolplasmas (infect animals, plants, insects), and entomoplasmas (infect insects, plants). The family Mycoplasmataceae contains two genera that infect animals: Mycoplasma and Ureaplasma, which are usually referred to collectively as mycoplasmas. There are many species of mycoplasmas, only four are recognized as human pathogens; *Mycoplasma pneumoniae*, *Mycoplasma hominis*, *Mycoplasma genitalium*, and *Ureaplasma urealyticum*. Although there are other species that have been isolated from humans but their role in disease is not well established. Members of the genus *Mycoplasma* are obligate parasites and pathogens of a wide range of mammalian and avian hosts and tend to be adapted to a specific host. Whilst this may result in a commensal type relationship, some species cause serious disease in domestic animals – for example contagious bovine pleuropneumonia (Mycoplasma mycoides subsp. *mycoides*). Phytoplasma are cell wall-less prokaryotes which infect plants. They exist exclusively in the phloem and are transmitted by grafting, dodder, vectors, insects and primarily leafhoppers. They cause symptoms such as yellowing, phyllody and witches' brooms in plants. The order Entomoplasmatales contains two families Entomoplasmataceae and Spiroplasmataceae. Spiroplasmas establish an array of interactions with plants and insects, ranging from epiphytic to symbiotic to pathogenic under family Spiroplasmataceae. While the *Entomoplasma* and *Mesoplasma* are included under family Entomoplasmataceae. The detailed informations about orders families and genera of class Mollicutes are discussed in second chapter.

Many characteristics used in bacterial taxonomy relate to cell wall and structures associated with it, such as flagellae, axial filaments, etc. One would expect that organisms without a cell wall like mollicutes would be difficult to characterize taxonomically. It became clear that many morphological adaptations (helicity, motility, attachment organelles, etc.) had occurred in Mollicutes even within the limitations imposed by a single bounding membrane. In time, various characters emerged that appeared to separate the emerging mollicute taxa. The genome sizes of the organisms appeared to be bimodally distributed. The observed sizes showed a clear separation of simple Mollicutes with small genomes clearly associated with vertebrate cells (genera Mycoplasma and *Ureaplasma*) from *Acholeplasma* species, which have larger genomes that were less clearly cell associated, some of which had been reported from sewage and soil. Because the latter organisms, unlike the *Mycoplasmas* and *Ureaplasmas*, did not require sterols in their culture media, the requirements of the more fastidious Mollicutes for sterol was added to phenotypic characters of mollicute. In some cases, unique metabolic pathways were discovered that characterized higher taxa. These included especially urease activity in members of the genus *Ureaplasma*, which are associated with the urinary tracts of vertebrate animals.

Evolution and Phylogeny of Mollicutes

The phylogeny of mollicutes is interesting as regards the degeneration of the genome. During their evolutionary history multiple reductions in genome size have occurred, the usual genetic code has been altered, and the overall rate of evolution uncharacteristically high. One suggested reason for the reduction in genome size is the evolution of mollicutes into strict parasites making much of their metabolic machinery obsolete. Mycoplasmas also have unique use of the amino acid codon UGA, which they use as an additional codon for tryptophan, while other organisms use it as a stop codon.

Due to degenerative or reductive evolution, accompanied by significant losses of genomic sequences, the genomes of mollicutes have shrunk and are relatively small compared to other bacteria, ranging from 580 kb. to 2,200 kb. The original indication that the higher taxa represented by the clusters of organisms sharing these properties would prove to be monophyletic was shown to be largely true by emerging molecular phylogenetic reconstructions. Earlier phylogenies and the Weisburg work indicated that genome size reductions must have occurred on at least two and possibly as many as four separate occasions in the course of mollicute phylogeny. Further, genome sizes of mollicutes, when examined carefully by pulsed field electrophoresis, proved to be far from bimodal and were in fact variable, even among members of monophyletic clusters. The collapse of genome size as a definitive character for mollicutes was followed soon after by a second major collapse. Studying the sterol requirements of a variety of mollicutes, and spurred by the phylogenetic reconstructions that suggested that spiroplasmas were relatively closely related to the acholeplasmas, other mollicutologists Rose and colleagues reexamined the sterol requirements of Spiroplasma species. They found that some of the faster-growing spiroplasmas were capable of multiplication through 23 consecutive passages in sterol-free media. Thus, a character that had been thought to define the genus turned out to be variable within the genus.

Although the revelation that genome size and sterol requirement were evolutionarily labile was unwelcome, most mollicutologists were nevertheless pleased that the group was essentially monophyletic and that it therefore fully merited the assignment of a procaryote division (Tenericutes) that had been given it in the proposal of Gibbons and Murray.

The phylogenetic constructions of Weisburg and colleagues greatly changed the course of mollicutology and alerted classification concepts based on phenological criteria needed revision. Further, the collapse of major phenotypic properties that had been a firm basis for classification for many years, coupled with the emergence of a plethora of insect and plant isolates, paved the way for a revision of classification for the higher taxa of Mollicutes. In 1993, Tully and colleagues presented a classification that was tailored to conform, in most of its details, to current phylogenetic reconstructions. In this classification, the

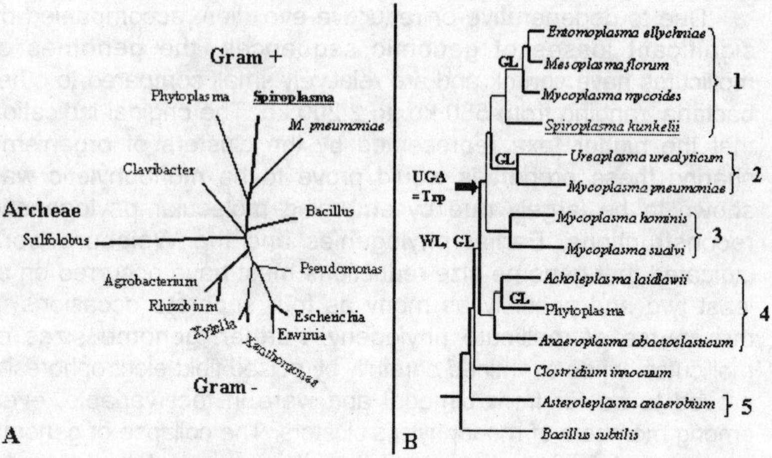

Fig. 1.1. Phylogenetic relationships of several bacterial clades containing bacterial pathogens (A), and the 5 phylogenetic groups within the Class Mollicutes (B). Plant pathogenic/symbiotic bacteria are indicated. GL, indicates gene loss, and WL, loss of cell wall.

indication that cultivable plant and insect Mollicutes had evolved monophyletically in close association with their hosts was recognized by construction of a new order, Entomoplasmatales, containing the Spiroplasmataceae and the Entomoplasmataceae. The latter family of nonhelical Mollicutes consisted of the nominate genus, which contained five species. These organisms failed to grow in serum-free medium, including many variations with Tween supplements or other combinations of single or mixed fatty acid supplements. A second genus of the Entomoplasmataceae, *Mesoplasma*, consisted of 10 species that could be grown in Tween-supplemented medium and that therefore presumably lacked a sterol requirement. All of the mesoplasmas and entomoplasmas could be distinguished from the vertebrate mycoplasmas by their temperature optima, which were always lower than 37°C and were often 32°C. The optimum for Mycoplasma species, most of which are associated with warm-blooded vertebrates, is always, or almost always, very close to 37°C. The exceptions concern isolates from cold-blooded vertebrates, such as alligators and tortoises, and a fish isolate. Also, none of the vertebrate mycoplasmas tested responded to Tween 80, providing an additional phenotypic character that separates the mesoplasmas from the mycoplasmas. In addition, the utility of 16S rDNA sequences for assessment of phylogenetic relationships above the species level has now been firmly established. Even at the species level, many or most species can be characterized by their 16S rDNA sequences.

Some characters may be useful for defining higher taxa and giving insight into higher level relationships. For example codon usage by spiroplasmas is similar to that of mycoplasmas, in that UGA is read as tryptophan. A large battery of metabolic pathways may characterize species clusters, genera, and families, but the universality of these characters has not been tested widely enough among species to establish their usefulness in practical taxonomy.

The type species of the genus *Mycoplasma* (*M. mycoides*) is clearly more closely related by 16S rDNA sequencing to the Entomoplasmataceae than to the Mycoplasmataceae. One possibility would be to change the generic name of the members of the *M. mycoides* species cluster. However, in the revised classification, it was necessary to retain this group in the Mycoplasmataceae because any change of the name of *M. mycoides* would be a major departure from provisions of the Bacteriological Code and would require an exception by the Judicial Commission. Further, change of the genus name would impose a significant taxonomic obstacle to workers studying a group of economically important organisms. One of the principal values of a classification system is the opportunity it provides for agricultural and public health officials and other nontaxonomists to identify the economically important organisms with some degree of certainty. Each time such a name

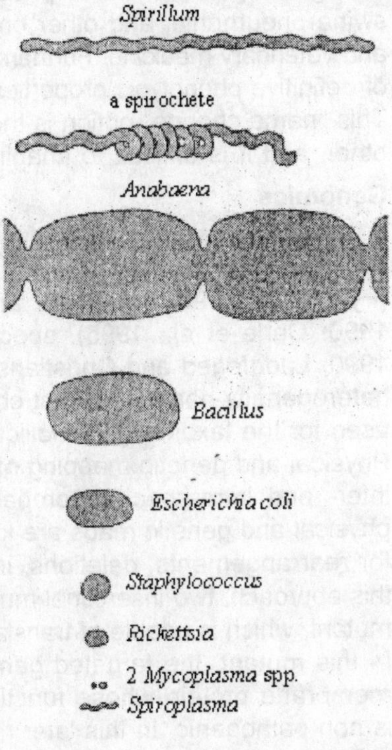

Fig. 1.2. Comparison of sizes of some eubacteria.

is changed, the value of taxonomy to the nontaxonomic community is diminished. Finally, it is not clear, even if the small clade of *M. mycoides* organisms was to be renamed, what phenotypic characters would be used for their identification. At present, there is no phenotypic character other than specific and extensive serological tests that could be used to separate these organisms from other species of the genus *Mycoplasma*. The alternative would be to rename the other (larger) group-the "true" mycoplasmas. However, this process would entail a name change for almost 100 species of vertebrate-associated organisms, including etiological agents of human pneumonia, postpartum fever, enzootic swine pneumonia, and other organisms of importance in human and veterinary medicine. Furthermore, the problems of the absence of definitive phenotypic properties of the two groups would remain. This "name change" option is therefore just as unworkable as the other, and it is difficult to imagine it ever being adopted.

Genomics

With the determination of genome sizes for many mollicutes by PFGE, a great heterogeneity has been observed at the genus (Pyle et al., 1988; Neimark and Lange, 1990; Whitley et al., 1990; Carle et al., 1995), species and strain level (Pyle et al., 1990; Ladefoged and Christiansen, 1992; Ye et al., 1995). This heterogeneity has meant that chromosome size alone cannot be used for the taxonomy of mollicutes (Neimark and Lange, 1990). Physical and genetic mapping of mollicute genomes has enabled inter- and intra-specific comparisons to be made. When both physical and genetic maps are known, genomes can be analysed for rearrangements, deletions, insertions or mutations. Following this approach, two insertional mutants were used. One is a motility mutant, which is unable of translational motility in viscous medium. In this mutant, the targeted gene encodes a highly hydrophobic membrane protein whose function is still unknown. The second is non pathogenic. In this later mutant the transposon is inserted into the first gene of the fructose operon, which encodes a regulatory protein. When experimentally transmitted to periwinkle by the leafhopper vector, this mutant multiplies in the host plant but does not induce symptoms. Further characterization of this mutant showed that the inability to induce symptoms was related to the inability to use fructose. For both mutants, functional complementation by transformation with the relevant wild-type DNA proved the disrupted locus to be involved in the observed phenotype. A classical approach to study gene function includes the following steps: isolation of mutants that are deficient for a given function, identification of the mutated gene, restoration of the phenotype by transformation of the mutant with the corresponding wild-type DNA, characterization of the gene product, and eventually determination of its function. *S. citri* mutants were produced by transformation with Tn4001 and were screened for both motility, on the basis of colony morphology, and for insect transmissibility and plant pathogenicity, by experimental transmission to periwinkle with the leafhopper vector Circulifer haematoceps.

Noncultivable Mollicutes

One of the major events in mollicute taxonomy during the past decade has been the use of 16S rDNA sequences to classify noncultivable wall-less procaryotes. The study of phytoplasmas has now reached increasingly high levels of sophistication. Recently, several other "noncultivable" wall-less disease agents of vertebrates have been characterized and assigned to the class Mollicutes. These include the agent of gray lung disease in rodents and an important group of hemotrophic organisms previously assigned to two genera in the rickettsiae (*Haemobartonella* and *Eperythrozoon*). Although new noncultivable species for which 16S rDNA sequence data provide evidence for uniqueness but that have not been thoroughly characterized must be given a "Candidatus" status, these blood parasites were given species names years ago. It is uncertain how their nomenclature will be changed to reflect their new status. A major problem is the existence of other species of *Eperythrozoon* and *Haemobartonella*, for which 16S rDNA data is not available.

It is often said that, in mollicutes, genetic studies have been hampered by the lack of suitable gene transfer systems. This is especially true for the nonculturable phytoplasmas. In the case of *S. citri*, the requirements for genetic analysis are being fulfilled. *S. citri* can be efficiently transformed with purified DNA by electroporation, insertional mutants can be obtained by transformation with transposon Tn4001 and plasmid vectors that offer the way to complement mutations have been constructed. However some limitations come from the fact that the molecular genetic tools that have been developped with a particular strain of *S. citri* cannot necessarily be used with others spiroplasmas species, even with different strains of the same species.

Plant Pathogenic Mollicutes

Plant pathogenic mollicutes are eubacteria associated with several hundred of plant diseases in fruit trees, grapevine, vegetables and ornamentals. As mollicutes, they share common properties with animal mycoplasmas such as the absence of a cell-wall and reduced genome size and G+C content. Phytopathogenic mollicutes range within two groups, spiroplasmas and phytoplasmas. Spiroplasmas are characterized by their helical morphology and are available in culture. Three of them, *Spiroplasma citri*, *Spiroplasma kunkelii* and *Spiroplasma phoeniceum* are plant pathogens. In contrast, the phytoplasmas are pleiomorphic and have never been cultured. The spiroplasmas represent a relatively "late" phylogenetic group with UGA coding for tryptophan whereas the phytoplasmas are part of the phylogenetically "early" acholeplasma branch with UGA being a stop codon. All phytopathogenic mollicutes, spiroplasmas as well as phytoplasmas, inhabit the phloem sieve tubes of plants and are transmitted from plant to plant by phloem-feeding insects, generally leafhoppers or psyllids.

Progress in detection, taxonomy, phylogeny, and characterization of plant mollicutes has been due essentially to the availability of monoclonal antibodies and to the development of the recombinant DNA technology. Because the phytoplasmas could not be cultured, the molecular techniques were applied first to the culturable spiroplasmas, in particular to *S. citri*, but they were soon applied to the non-culturable phytoplasmas. In these organisms, breakthroughs were the purification of phytoplasmal DNA (based on its high AT content) from infected insects or plant tissues, and the use of pulse-field electrophoresis to obtain full-length phytoplasmal chromosome free of contaminant plant or insect DNA. More recently, monoclonal antibodies were used to screen genomic DNA libraries and to purify membrane protein by immunoaffinity chromatography, leading to the characterization of genes encoding major surface proteins. Spiroplasmal as well as phytoplasmal genomes have been sized and mapped, many genes have been sequenced, and proteins were characterized. Also, in the near future, the complete genome sequences of plant pathogenic mollicutes, including *S. citri*, will be available. However, in spite of this considerable amount of information, very few is known about the genetic systems which control the biology of these organisms. The fact that spiroplasmas could be cultured in vitro, whereas the phytoplasmas could not, has made *S. citri* a privileged model for molecular genetic studies of these phloem-restricted plant pathogens. Genes can be identified by analysis of their mutations, whether spontaneous or induced. However analysis of gene mutations requires that genes be transferred between strains. In the recent years, efforts have been made to develop gene transfer systems, including electrotransformation, construction of plasmid vectors, and insertional mutagenesis, in order to identify genes and molecular mechanisms that govern the interactions of the helical mollicute *S. citri* with its two eucaryotic hosts, the leafhopper vector, and the plant.

International Organization for Mycoplasmology (IOM)

Mollicutes dramatically differ from other bacteria by their minute size (0.3 - 0.5 µm), total lack of cell wall and small genomes. The first efforts toward the study of those microbial forms that would eventually be called mollicutes really began with the pioneering efforts of a group of French bacteriologists. In 1898, E. Nocard and E.R. Roux, and their collaborators, A. Borrel, A. Salimbeni, and E. Dujardin-Baumetz, first cultured the microbe of contagious pleuropneumonia of cattle on artificial broth. Two years later, E. Dujardin-Baumetz was able to adapt the organism to a solid medium. He observed the small colonies with the now familiar dark center and light peripheral area, and documented the filterability of the organism. Twenty-five years went by (1923) before J. Bridre and A. Donatien, again in France, cultivated the organism associated with agalactia of sheep and goats and established the etiologic role of the organism in the disease.

These discoveries stimulated considerable interest in filterable bacteria and by the mid to late 1930s more researchers began to study additional members of this group of microbes that had come to be known as "pleuropneumonia-like organisms," or "PPLO". In England, Ledingham worked on their morphology; Klieneberger-Nobel was isolating the organisms from rodents; and Laidlaw and Elford were cultivating what would later turn out to be acholeplasmas from sewage. In Germany, Seiffert was also growing so-called saprophytic PPLO from soil and compost. In the US, Nelson was isolating the so-called pleuropneumonia-like organisms from chickens and rodents, Sabin was also studying similar organisms with neurotoxic properties, and, in 1937, Dienes and Edsall cultivated the first pleuropneumonia-like organism from the urogenital tract of a female patient.

Although the period of 1940-60 saw further progress in understanding the biology of these organisms, the 1962 discovery of the causative agent of primary atypical pneumonia of humans provided not only an important stimulus to study further the organisms in man and their pathogenicity, but it provoked a major change in the classification and nomenclature of the organisms. Now these filterable, wall-less prokaryotes were given the collective term "mycoplasmas," since it was proposed by an international taxonomy subcommittee that all organisms in this collection be assigned to the genus Mycoplasma. However, this terminology became outdated relatively soon, because many new and different mycoplasmas were being isolated and characterized, and new taxonomic names and classification levels were being used to separate these organisms from those in the genus Mycoplasma (such as members of the genus Acholeplasma, genus Ureaplasma, etc.). Eventually, a revised classification scheme was devised whereby all filterable, wall-less prokaryotes were covered under a broad system called the class Mollicutes.

One might well consider the next thirty-years as the "golden age" of mycoplasmology. Major advances occurred on many fronts, including the discovery of mollicutes in plant and insect hosts, the first helical mollicutes (spiroplasmas), new strictly anaerobic mollicutes (anaeroplasmas), of mycoplasma and acholeplasma viruses, the occurrence of many newly isolated mollicutes from man and animals and their pathogenicity, the expanding field of molecular biology and genetics of mollicutes, and new concepts of how mollicutes interact with host immune responses. Some of the most exciting developments are now taking place in understanding how mollicutes might avoid host immune defenses and interact with other microbial agents to induce disease. Again, these major advancements have produced changes in our understanding of the evolution and classification of Mollicutes, so that we now recognize such new organisms as the mesoplasmas, entomoplasmas, asteroleplasmas, etc.

Events leading to the eventual formation of an international society of workers in the field of mycoplasmology really began in early 1972. The Ciba Foundation in London sponsored a symposium on "Pathogenic Mycoplasmas", whereby a small group of researchers interested in human, animal, plant, and insect mollicute diseases gathered to discuss problems of mutual interest. This symposium stimulated considerable international interaction and cooperation among workers in the field, and eventually led to discussions and plans for a larger international symposium on "Mycoplasmas of Man, Animals, Plants and Insects" in Bordeaux, France in 1974. Over 300 participants attended this meeting, sponsored by the University of Bordeaux II and the Institut National de la Recherche Agronomique (INRA) in France, from 31 countries. The proceedings of the conference were published as a document in the Institut National de la Santé et de la Recherche Médicale (INSERM) series (volume 33).

During the Bordeaux symposium, discussions were held with some of the participants regarding whether an international organization might be formed to stimulate further cooperation and promote future meetings. As a result of these discussions, a group of individuals were invited to meet (including D.G. Edward, E.A. Freundt, G.S. Cottew, J.M. Bove, W. Bredt, M.F. Barile, S. Razin, R.F. Whitcomb, and J.G. Tully) to plan a series of ad hoc committees for the formation of a new international organization. The committees were charged with the preparation of a constitution for the organization, to nominate candidates for office, to formulate membership publicity and financial appeals, and to plan for the first scientific congress of the organization. As a consequence of these joint efforts, the first election of officers for the newly formed International Organization for Mycoplasmology occurred in the Spring of 1976: Chair, Joseph G. Tully (USA); Chair-Elect, Shmuel Razin (Israel); Secretary-General, D. Taylor-Robinson (UK), and Treasurer, Michael F. Barile (USA). At Large Members appointed to the Board included: G. Biberfeld (Sweden), J.M. Bove (France), E.A. Freundt (Denmark), L. Stipkovits (Hungary), and R.F. Whitcomb (USA). The numerous individuals and members of the IOM who have later served the Organization in elective office or in appointed positions and have contributed greatly to the growth and recognition of the Organization are listed in the IOM Handbook.

The First International Congress of the IOM was held in September 1976 in Glasgow, Scotland, in conjunction with a meeting of the British Society for General Microbiology. D. Taylor-

Robinson was the Chair of the Program Committee for the meeting. Over 90 communications were presented during the congress, and the abstracts were published in the Proceedings of the Society for General Microbiology (Vol. III, Part 4, August 1976). The Second International Congress was held August 28 to September 1, 1978 in Freiburg, Germany, and W. Bredt was the Chair of the Program Committee. Abstracts of this meeting were published in Zentralblatt fur Bakteriologie, Parasitenkunde, Infektions-krankheiten und Hygiene (Band 241 (2), 1978). The Third International Congress of IOM took place in Custer, South Dakota, USA, on September 3-9, 1980. The Chair of the Program Committee was P.F. Smith, and the proceedings of the meeting were published in Reviews of Infectious Diseases (Vol.4, supplement, 1982). The Fourth Congress of IOM occurred on September 1-7, 1982 in Tokyo, Japan. O. Kitamoto was the Chair of the Program Committee, with the proceedings published in the Yale Journal of Biology and Medicine (Vol. 56, Nos. 5-6, 1983). The Fifth Congress of IOM was held in Jerusalem, Israel from June 24-29, 1984, with S. Razin as the Chair of the Program Committee. Symposia papers were published in the Israel Journal of Medical Sciences (Vol. 20, Nos.9-10, 1984). The Sixth IOM Congress was held on August 26-30, 1986 in Birmingham, Alabama, with G.H. Cassell as Chair of the Program Committee. Proceedings were published in the Israel Journal of Medical Sciences (Vol. 23, 1987). The Seventh IOM Congress was held on June 2-9, 1988 in Baden (near Vienna), Austria. The Chair for the Program Committee was G. Stanek and the proceedings appeared in a supplement to Zentralblatt für Bakteriologie, Mikrobiologie und Hygiene. The Eighth IOM Congress took place in Istanbul, Turkey July 8-12, 1990, with S. Rottem as Program Chair. Meeting abstracts appeared in IOM Letters, vol.1. The Ninth IOM Congress convened in Ames, Iowa on August 2-7, 1992, with R. Rosenbusch as Program Chair. Meeting abstracts appeared in IOM Letters, vol. 2. The Tenth IOM Congress, which represented the 20-year anniversary of the rounding of IOM, took place on July 19-26, 1994 in Bordeaux, France. It was significant and highly appropriate that this milestone in the history and development of IOM occurred in the city where the concept of an international mycoplasmology society was born. The Eleventh IOM Congress took place in Orlando, Florida on July 14-19, 1996, with J. K.

Davis as Program Chair. Meeting abstracts appeared in IOM Letters, vol. 4. During the past years, the IOM has either set aside funds or has had the cooperation and generosity of corporate and governmental sponsors in funding student travel fellowships to IOM meetings. This is an important contribution to developing interest and training within the field for new investigators.

It is appropriate also to detail here some of the other major events that have occurred within the development of IOM and how these have contributed to the field of mycoplasmology. In 1978-79, two important additions occurred in the area of IOM international activities. The International Research Program on Comparative Mycoplasmology (IRPCM), which had been launched in 1971 under auspices of the World Health Organization (WHO) and the Food and Agricultural Organization (FAO) of the United Nations, became an integral part of the IOM. This collaborative research program was initially formed to promote the characterization of animal mycoplasmas at the international level, and it operated through a directing group (the Board) composed of the heads of specific working teams and a group of outside experts. The Board and program coordinated information on animal mycoplasmas and their diseases, and published a series of working group documents on standard mycoplasma techniques. Today, the IRPCM is composed of mycoplasmologists from over 100 laboratories around the world collaborating within thirteen working teams on the development of improved techniques for isolation and characterization of mollicutes from man, animals, plants and insects. In 1983, the IRPCM sponsored publication of the first two volumes of Methods in Mycoplasmology (Academic Press, Inc.), providing wide dissemination of standard techniques for mollicutes.

In 1979, in association with the University of Bordeaux II and the French Institut National de la Recherche Agronomique, the IOM began the sponsorship of the first of a series of International Mycoplasma Techniques Courses in Bordeaux, France. Twenty-nine participants from 16 countries spent three weeks in an intensive and comprehensive program learning basic bench techniques for mollicutes. The faculty was composed of 15 IOM members from outside France and 5 local faculty. A repeat of the course was given in 1983, with 31 participants from 14 countries. In 1987, the central theme of the course was changed to an emphasis on gene technology and the molecular biology of mollicutes, including monoclonal antibody techniques. Thirty students, from 9 countries, attended the two and 1/2 week course in Bordeaux. Again, eleven IOM members from Bordeaux and outside areas participated in the teaching program. In collaboration with the Federation of European Microbiological Societies (FEMS), the IOM also participated in a techniques course on "Rapid Diagnosis of Mycoplasmas" at the Hebrew University, Hadassah Medical School in Jerusalem, Israel on August 11-23, 1991. Again, in all of these programs, the IOM has contributed funds for fellowships to cover partial support of travel expenses of student participants.

In addition to the biennial congresses, the IOM has sponsored a number of special symposia. The first, "Mycoplasma hominis, a Human Pathogen" was organized by P-A. Mardh (Sweden), B.R. Moller (Denmark), and W.M. McCormack (USA), and was co-sponsored with a number of other organizations. The symposium was held March 8-12, 1983 in Bietostolen, Norway, with the proceedings published in Sexually Transmitted Diseases (Vol. 1 O, supplement, 1983). The second symposium, entitled "Ureaplasmas of Humans, with Emphasis Upon Maternal and Neonatal Infections" was held in Seattle, Washington (USA) on October 10-12, 1985. The organizers were G.H. Cassell (USA), W.A. Clyde (USA), G.E. Kenny (USA), W.M. McCormack (USA), and D. Taylor-Robinson (UK). Abbott Laboratories acted as co-sponsor. In December 1991, the IOM cosponsored, along with The National Institute of Allergy and Infectious Diseases and Abbott Laboratories, an international symposium in Scottsdale, Arizona on "The Changing Role of Mycoplasmas in Respiratory Disease and AIDS". The Program Chair was G.H. Cassell, with a Program and Planning Committee of J.B. Baseman, J.M. Bove, S.C. Lo, L. Montagnier, R.L.Quackenbush, D. Taylor-Robinson, and J.G. Tully. The proceedings were published in Clinical Infectious Diseases 17 (supplement 1), in August 1993.

Today, the IOM represents a growing membership of active research scientists, teachers, and industrial and commercial scientists, including a broad spectrum of scientific disciplines within microbiology, biochemistry, molecular biology, genetics, immunology, virology, clinical medicine, veterinary science, plant pathology, and entomology. The members, from all parts of the world, have through their support of the IOM promoted cooperative international study and dissemination of knowledge of wall-less eubacteria and the diseases they cause. We look forward to the second "golden age" in mycoplasmology in the years to come.

During the past decade, research has yielded new knowledge about the plant and insect host ranges, geographical distribution, and phylogenetic relationships of phytoplasmas, and a taxonomic system has emerged in which distinct phytoplasmas are named as separate "Candidatus phytoplasma species." In large part, this progress has resulted from the development and use of molecular methods to detect, identify, and classify phytoplasmas. While these advances continue, research has recently begun on the phytoplasma genome, how phytoplasmas cause disease, the role of mixed phytoplasmal infections in plant diseases, and molecular/genetic phenomena that underlie symptom development in plants. These and other recent advances are laying the foundation for future progress in understanding the mechanisms of phytoplasma pathogenicity, organization of the phytoplasma genome, evolution of new phytoplasma strains and emergence of new diseases, bases of insect transmissibility and specificity of transmission, and plant gene expression in response to phytoplasmal infection, as well as the design of novel approaches to achieve effective control of phytoplasmal diseases.

2

Classification of Mollicutes

Taxonomy of Mollicute

Microorganisms classified as Mollicutes (Procaryotae: Tenericutes: Mollicutes) may be the smallest living cellular organisms. This property focused the attention of the general biological community on the group in the late 1960s and stimulated a discussion of the minimal genetic and biochemical traits that the smallest known free-living organism must possess. This interest has persisted to the present and has led to the complete sequencing of the genomes of two Mycoplasma species. In early studies, most Mollicutes assigned to the genus Mycoplasma appeared to be exclusively host associated, particularly with vertebrates. However, some of the organisms isolated during this period did not require sterols or cholesterol for growth. This distinction, as well as the fact that some of the organisms had a genome size twice the size of Mycoplasma species, prompted a proposal to designate the genus Acholeplasma and, later, to elevate it to ordinal rank.

The taxonomic characterization and description of *Spiroplasma citri* in 1973 posed certain classification problems in the higher taxonomic levels of the Mollicutes. The first helical, sterol-requiring mollicute of plant and/or arthropod origin was clearly different from all other known Mollicutes. Other newly isolated spiroplasmas described shortly thereafter had similar growth characteristics and also appeared to possess a sterol requirement, but their genome sizes were more in the size range of the acholeplasmas. Eventually, the genus Spiroplasma was elevated to a family within the order Mycoplasmatales. The isolation of a variety of other plant and insect Mollicutes in the late 1970s to the early 1980s brought further complications to mollicute taxonomy. The non-helical strains obtained from flowers and other plant surfaces and from the gut and hemolymph of numerous insects collected over a wide geographic region in the United States by Clark and Whitcomb, appeared to be very similar to the acholeplasmas but could not be grown in serum-free media and had genome sizes much smaller than classic acholeplasmas. These organisms were eventually placed in Mesoplasma or Entomoplasma.

Attempts to properly place the emerging taxa of Mollicutes eventually led to the landmark study of Weisburg and colleagues whereby 16S rDNA sequences of almost 50 mollicute species were examined. When information from this study became available, it appeared that the Mollicutes had descended, presumably by a monophyletic route, from gram positive bacteria with a consequent reduction in their genome size from values as high as 2,800 kbp, typical of such genera as Streptococcus and Lactobacillus and some clostridia, to about 500-800 kpb to 1000-2000 kbp. These phylogenetic studies eventually culminated in a revision of higher level taxa of Mollicutes .

Mollicute classification cannot entirely mirror phylogeny. The existence of a phylogenetic classification of Mollicutes led to a more satisfactory classification than previous versions based on labile phenotypic characters, and updated revisions in the individual phylogenetic clusters have begun to appear as more sequence analyses are undertaken. The new classification has led to hypotheses about phenotypic characters, which accelerated the rate of discovery of metabolic diversity in the Mollicutes. Nevertheless, as appreciative as mollicute workers have become of the phylogenetic classification, it has proved impossible to utilize phylogenetic position in every aspect of the classification. A cluster of *Mycoplasma* species, including the type species (*M. mycoides*) of the order Mycoplasmatales, family Mycoplasmataceae, and genus Mycoplasma, appeared to be divided from the rest of the genus by a deep branch and, indeed, to be more closely aligned with the Spiroplasmataceae than the Mycoplasmataceae. Even more unexpected, this group appeared to cluster rather closely with a group of nonhelical Mollicutes from plants and insects. Some of the members of this group (now mesoplasmas) were able to grow in serum-free medium supplemented with 0.04% Tween 80, while others (now entomoplasmas) had a sterol requirement. The variable sterol requirements of these isolates had prompted an earlier proposal to assign some to the genus Mycoplasma and some to the genus Acholeplasma.

Taxa Covered by the ICSP Subcommittee on the Taxonomy of Mollicutes - June 2003

The class Mollicutes, which is currently divided into three orders (Mycoplasmataceae, Acholeplasmataceae, Spiroplasmataceae); each order contains different families, and six genera are distinguished by classical taxonomy. For instance, the genus Mycoplasma comprises more than 50 known species, which are further subdivided into a multitude of strains. The following list of mollecutes is based in part on Bergey's Manual of Systematic Bacteriology, 2nd edition (Springer, New York). It is prepared by Joseph G. Tully and Janet M. Bradbury and updated by them on 6 June 2003.

Class—Mollecutes

Order—Mycoplasmatales

Family Mycoplasmataceae—The family Mycoplasmataceae contains two genera that infect humans: Mycoplasma and Ureaplasma, which are usually referred to collectively as mycoplasmas.

Genus—Mycoplasma (Type Genus)

Mycoplasma mycoides subsp. mycoides (type species)	Typestrain:PG1
Mycoplasma mycoides subsp. capri	Type strain: PG3
Mycoplasma adleri	Type strain: G145
Mycoplasma agalactiae	Type strain: PG2
Mycoplasma agassizii	Type strain: PS6
Mycoplasma alkalescens	Type strain:D12(PG51)
Mycoplasma alligatoris	Type strain A21JP2
Mycoplasma alvi	Type strain Ilsley
Mycoplasma anatis	Type strain. 1340
Mycoplasma anseris	Type strain: 1219
Mycoplasma arginini	Type strain: G230
Mycoplasma arthritidis	Type strain: PG6

Mycoplasma auris	Type strain: U1A
Mycoplasma bovigenitalium	Type strain: PG11
Mycoplasma bovirhinis	Type strain: PG43
Mycoplasma bovis	Type strain: Donetta (PG45)
Mycoplasma bovoculi	Type strain: M165/69
Mycoplasma buccale	Type strain: CH20247
Mycoplasma buteonis	Type strain: Bb/T2g
Mycoplasma californicum	Type strain: ST-6
Mycoplasma canadense	Type strain: 275C
Mycoplasma canis	Type strain: PG14
Mycoplasma capricolum subsp. capricolum	Type strain: California kid
Mycoplasma capricolum subsp. capripneumoniae	Type strain: F38
Mycoplasma caviae	Type strain: G122
Mycoplasma cavipharyngis	Type strain: 117C
Mycoplasma citelli	Type strain: RG-2C
Mycoplasma cloacale	Type strain: 383
Mycoplasma collis	Type strain: 58B
Mycoplasma columbinasale	Type strain: 694
Mycoplasma columbinum	Type strain: MMP1
Mycoplasma columborale	Type strain: MMP4
Mycoplasma conjunctivae	Type strain: HRC581
Mycoplasma corogypsi	Type strain: BV1
Mycoplasma cottewii	Type strain: VIS
Mycoplasma cricetuli	Type strain: CH
Mycoplasma crocodyli	Type strain: MP145
Mycoplasma cynos	Type strain: H831
Mycoplasma dispar	Type strain: 462/2
Mycoplasma edwardii	Type strain: PG24
Mycoplasma elephantis	Type strain: E42
Mycoplasma equigenitalium	Type strain: T37
Mycoplasma equirhinis	Type strain: M432/72
Mycoplasma falconis	Type strain: H/T1
Mycoplasma fastidiosum	Type strain: 4822
Mycoplasma faucium	Type strain: DC-333
Mycoplasma felifaucium	Type strain: PU

Mycoplasma feliminutum	Type strain: Ben
Mycoplasma felis	Type strain: CO
Mycoplasma fermentans	Type strain: PG18
Mycoplasma flocculare	Type strain: Ms42
Mycoplasma gallinaceum	Type strain: DD
Mycoplasma gallinarum	Type strain: PG16
Mycoplasma gallisepticum	Type strain: PG31(X95)
Mycoplasma gallopavonis	Type strain: WR1
Mycoplasma gateae	Type strain: CS
Mycoplasma genitalium	Type strain: G37
Mycoplasma glycophilum	Type strain: 486
Mycoplasma gypis	Type strain: B1/T1
Mycoplasma haemocanis	Type strain: not established
Mycoplasma haemofelis	Type strain: not established
Mycoplasma haemomuris	Type strain: not established
Mycoplasma hominis	Type strain: PG21
Mycoplasma hyopharyngis	Type strain: H3-6BF
Mycoplasma hyopneumoniae	Type strain: J
Mycoplasma hyorhinis	Type strain: BTS-7
Mycoplasma hyosynoviae	Type strain: S16
Mycoplasma imitans	Type strain: 4229
Mycoplasma indiense	Type strain: 3T
Mycoplasma iners	Type strain: PG30
Mycoplasma iowae	Type strain: 695
Mycoplasma lagogenitalium	Type strain: 12MS
Mycoplasma leonicaptivi	Type strain: 3L2
Mycoplasma leopharyngis	Type strain: LL2
Mycoplasma lipofaciens	Type strain: R171
Mycoplasma lipophilum	Type strain: MaBy
Mycoplasma maculosum	Type strain: PG15
Mycoplasma meleagridis	Type strain: 17529
Mycoplasma microti	Type strain: IL371
Mycoplasma moatsii	Type strain: MK 405
Mycoplasma mobile	Type strain: 163K
Mycoplasma molare	Type strain: H542
Mycoplasma muris	Type strain: RIII-4
Mycoplasma mustelae	Type strain: MX9

Mycoplasma neurolyticum	Type strain: Type A
Mycoplasma opalescens	Type strain: MH5408
Mycoplasma orale	Type strain: CH 19299
Mycoplasma ovipneumoniae	Type strain: Y98
Mycoplasma oxoniensis	Type strain: 128
Mycoplasma penetrans	Type strain: GTU-54-6A1
Mycoplasma phocacerebrale	Type strain: 1049
Mycoplasma phocarhinis	Type strain: 852
Mycoplasma phocidae	Type strain: 105
Mycoplasma pirum	Type strain: 70-159
Mycoplasma pneumoniae	Type strain: FH
Mycoplasma primatum	Type strain: HRC292
Mycoplasma pullorum	Type strain: CKK
Mycoplasma pulmonis	Type strain: Ash (PG34)
Mycoplasma putrefaciens	Type strain: KS1
Mycoplasma salivarium	Type strain: PG20(H110)
Mycoplasma simbae	Type strain: LX
Mycoplasma spermatophilum	Type strain: AH159
Mycoplasma spumans	Type strain: PG13
Mycoplasma sturni	Type strain: UCMF
Mycoplasma sualvi	Type strain: Mayfield B
Mycoplasma subdolum	Type strain: TB
Mycoplasma suis	Type strain: not established
Mycoplasma synoviae	Type strain: WVU 1853
Mycoplasma testudinis	Type strain: 01008
Mycoplasma verecundum	Type strain: 107
Mycoplasma wenyonii	Type strain: not established
Mycoplasma yeatsii	Type strain: GIH
Genus II—Ureaplasma	
Ureaplasma urealyticum (type species)	Type strain: T960(CX8)
Ureaplasma canigenitalium	Type strain: D6P-C
Ureaplasma cati	Type strain: F2
Ureaplasma diversum	Type strain: A417
Ureaplasma felinum	Type strain: FT2-B
Ureaplasma gallorale	Type strain: D6-1
Ureaplasma parvum	Type strain: 27

Order—Entomoplasmatales

Family—Entomoplasmataceae

Genus—Entomoplasma (type genus)

Entomoplasma ellychniae (type species)	Type strain: ELCN-1
Entomoplasma freundtii	Type strain: BARC 318
Entomoplasma lucivorax	Type strain: PIPN-2
Entomoplasma luminosum	Type strain: PIMN-1
Entomoplasma melaleucae	Type strain: M1
Entomoplasma somnilux	Type strain: PYAN-1

Genus II—Mesoplasma

Mesoplasma florum (type species)	Type strain: L1
Mesoplasma chauliocola	Type strain: CHPA-2
Mesoplasma coleopterae	Type strain: BARC 779
Mesoplasma corruscae	Type strain: ELCA-2
Mesoplasma entomophilum	Type strain: TAC
Mesoplasma grammopterae	Type strain: GRUA-1
Mesoplasma lactucae	Type strain: 831-C4
Mesoplasma photuris	Type strain: PUPA-2
Mesoplasma pleciae	Type strain: PS-1
Mesoplasma seiffertii	Type strain: F7
Mesoplasma syrphidae	Type strain: YJS
Mesoplasma tabanidae	Type strain: BARC 857

Family—Spiroplasmataceae

Genus—Spiroplasma (type genus)

Spiroplasma citri (type species)	Group I-1.	Type strain: R8A2
Spiroplasma melliferum	Group I-2.	Type strain: BC-3
Spiroplasma kunkelii	Group I-3.	Type strain: E275
Spiroplasma insolitum	Group I-6.	Type strain: M55
Spiroplasma phoeniceum	Group I-8.	Type strain: P40
Spiroplasma poulsonii	Group II.	Type strain: DW-1
Spiroplasma floricola	Group III.	Type strain: OBMG
Spiroplasma apis	Group IV.	Type strain: B31
Spiroplasma mirum	Group V.	Type strain: SMCA
Spiroplasma ixodetis	Group VI.	Type strain: Y32

Spiroplasma monobiae	Group VII.	Type strain: MQ-1
Spiroplasma syrphidicola	Group VIII-1.	Type strain: EA-1
Spiroplasma chrysopicola	Group VIII-2.	Type strain: DF-1
Spiroplasma clarkii	Group IX.	Type strain: CN-5
Spiroplasma culicicola	Group X.	Type strain: AES-1
Spiroplasma velocicrescens	Group XI.	Type strain: MQ-4
Spiroplasma diabroticae	Group XII.	Type strain: DU-1
Spiroplasma saubaudiense	Group XIII.	Type strain: AR-1343
Spiroplasma corruscae	Group XIV.	Type strain: EC-1
Spiroplasma cantharicola	Group XVI-1.	Type strain: CC-1
Spiroplasma turonicum	Group XVII.	Type strain: TAB4c
Spiroplasma litorale	Group XVIII.	Type strain: TN-1
Spiroplasma lampyridicola	Group XIX.	Type strain: PUP-1
Spiroplasma leptinotarsae	Group XX.	Type strain: LD-1
Spiroplasma taiwanense	Group XXII.	Type strain: CT-1
Spiroplasma gladiatoris	Group XXIII.	Type strain: TG-1
Spiroplasma chinense	Group XXIV.	Type strain: CCH
Spiroplasma diminutum	Group XXV.	Type strain: CUAS-1
Spiroplasma alleghenense	Group XXVI.	Type strain: PLHS-1
Spiroplasma lineolae	Group XXVII.	Type strain: TALS-1
Spiroplasma platyhelix	Group XXVIII.	Type strain: PALS-1
Spiroplasma montanense	Group XXXI.	Type strain: HYOS-1
Spiroplasma helicoides	Group XXXII.	Type strain: TABS-1
Spiroplasma tabanidicola	Group XXXIII.	Type strain: TAUS-1
Unnamed groups		
Spiroplasma sp.	Group I-4.	Type strain: 277 F
Spiroplasma sp.	Group I-5.	Type strain: LB-12
Spiroplasma sp.	Group I-7.	Type strain: N525
Spiroplasma sp.	Group VIII-3.	Type strain: TAAS-1
Spiroplasma sp.	Group XV.	Type strain: I-25
Spiroplasma sp.	Group XVI-2.	Type strain: CB-1
Spiroplasma sp.	Group XVI-3.	Type strain: AR-1357
Spiroplasma sp.	Group XXI.	Type strain: W-115
Spiroplasma sp.	Group XXIX.	Type strain: TIUS-1
Spiroplasma sp.	Group XXX.	Type strain: BIUS-1
Spiroplasma sp.	Group XXXIV.	Type strain: B1901

Order—Acholeplasmatales

Family—Acholeplasmataceae

Genus—Acholeplasma (type genus)

Acholeplasma laidlawii (type species)	Type strain: PG8
Acholeplasma axanthum	Type strain: S-743
Acholeplasma brassicae	Type strain: 0502
Acholeplasma cavigenitalium	Type strain: GP3
Acholeplasma equifetale	Type strain: C112
Acholeplasma granularum	Type strain: BTS39
Acholeplasma hippikon	Type strain: C1
Acholeplasma modicum	Type strain: PG49
Acholeplasma morum	Type strain. 72-043
Acholeplasma multilocale	Type strain: PN525
Acholeplasma oculi	Type strain: 19-L
Acholeplasma palmae	Type strain: J-233
Acholeplasma parvum	Type strain: H23M
Acholeplasma vituli	Type strain: FC 097-2

Order—Anaeroplasmatales

Family—Anaeroplasmataceae

Genus—Anaeroplasma (type genus)

Anaeroplasma abactoclasticum (type species)	Type strain: 6-1
Anaeroplasma bactoclasticum	Type strain: JR
Anaeroplasma intermedium	Type strain: 7LA
Anaeroplasma varium	Type strain: A-2

Genus—Asteroleplasma

Asteroleplasma anaerobium (type species)	Type strain: L1

Candidatus within the class Mollicutes

Candidatus within the genus Mycoplasma

'Candidatus Mycoplasma ravipulmonis' Neimark et al. (1998). Int J Syst Bacteriol 48, 389-394.

'Candidatus Mycoplasma haemominutum' Foley & Pedersen (2001). Int J Syst Evol Microbiol 51, 815-817.

'Candidatus Mycoplasma haemodidelphidis' Messick et al. (2002). Int J Syst Evol Microbiol 52, 693-698.

'Candidatus Mycoplasma haemolamae' Messick et al. (2002). Int J Syst Evol Microbiol 52, 693-698.

'Candidatus Mycoplasma haemocanis' Messick et al. (2002). Int J Syst Evol Microbiol 52, 693-698.

Candidatus within the genus Phytoplasma

'Candidatus Phytoplasma aurantifolia' Zreik et al. (1995). Int J Syst Bacteriol 45, 449-455.

'Candidatus Phytoplasma australasia' White et al. (1998). Int J Syst Bacteriol 48, 941-951.

'Candidatus Phytoplasma australiense' Davis et al. (1997). Int J Syst Bacteriol 47, 262-269.

'Candidatus Phytoplasma brasiliense' Montano et al. (2000). Int J Syst Evol Microbiol 51, 1109-1118.

'Candidatus Phytoplasma castaneae' Jung et al. (2002). Int J Syst Evol Microbiol 52, 1543-1549.

'Candidatus Phytoplasma fraxini' Griffiths et al. (1999). Int J Syst Bacteriol 49, 1605-1614.

'Candidatus Phytoplasma japonicum' Sawayanagi et al. (1999). Int J Syst Bacteriol 49, 1275-1285.

'Candidatus Phytoplasma phoeniceum' Verdin et al. (2003). Int J Syst Evol Microbiol 53, 833-838.

The NCBI Taxonomy for Nomenclature or Classification of Mollecutes

Class—Mollicutes

Order—Acholeplasmatales
- Family—Acholeplasmataceae
  - Genus—Acholeplasma
  - Genus—Phytoplasma (plant yellows agents)

Order—Anaeroplasmatales
- Family—Anaeroplasmataceae
  - Genus—Anaeroplasma
    - Asteroleplasma
- Genera incertae sedis
- Erysipelotrichaceae

Order—Entomoplasmatales
- Family—Entomoplasmataceae
  - Genus—Entomoplasma

Genus—Mesoplasma

Family—Spiroplasmataceae

Genus—Spiroplasma

Unclassified Spiroplasmataceae

Environmental samples

Uncultured Entomoplasmatales bacterium

Order—Mycoplasmatales

Family—Mycoplasmataceae

Genus—Mycoplasma

Genus—Ureaplasma

Unclassified Mollicutes

Unclassified Mollicutes (miscellaneous)

Clostridium innocuum

Clostridium aff. innocuum CM970

Clostridium ramosum

Clostridium spiroforme

Eubacterium biforme

Eubacterium cylindroides

Eubacterium dolichum

Eubacterium tortuosum

Lactobacillus catenaformis

Lactobacillus vitulinus

Mollicutes bacterium pACH93

Streptococcus pleomorphus

Clostridium sp. S9

Environmental samples

Unidentified mollicute

Unidentified oil field bacterium CLOS1

Order—Acholeplasmatales

Family—Acholeplasmataceae

Genus—Acholeplasma

Acholeplasma axanthum

Acholeplasma brassicae

Acholeplasma laidlawii

Acholeplasma modicum

Acholeplasma oculi

Acholeplasma palmae
Acholeplasma vituli
Acholeplasma sp.

Phytoplasma (plant yellows agents) 16SrI (Aster yellows group)

Aconitum proliferation phytoplasma
Alfalfa stunt phytoplasma
Apricot chlorotic leafroll phytoplasma
Aster yellows phytoplasma
Aster yellows phytoplasma A
Aster yellows phytoplasma B
Aster yellows phytoplasma O
Phaseolus vulgaris phytoplasma
Blueberry stunt phytoplasma
Candidatus Phytoplasma japonicum
Cherry little leaf phytoplasma
Cirsium yellows phytoplasma
Clover phyllody phytoplasma
Cuban sugarcane yellow leaf phytoplasma
Epilobium phyllody phytoplasma
Gerbera jamesonii phytoplasma
Ipomoea obscura witches'-broom phytoplasma
Lactuca sativa witches'-broom phytoplasma
Maize bushy stunt phytoplasma
Marigold phyllody phytoplasma
Mulberry dwarf phytoplasma
Oat proliferation phytoplasma
Onion yellows phytoplasma
Paulownia witches'-broom phytoplasma
Petunia flat stem phytoplasma
Potato purple top phytoplasma
Rose cv. Frisco phytoplasma
Strawberry phylloid fruit phytoplasma
Tomato big bud phytoplasma
Valeriana yellows phytoplasma
Phytoplasma sp. AY1
Phytoplasma sp. DIV
Phytoplasma sp. PY3

16SrII (Peanut WB group)
- Candidatus Phytoplasma aurantifolia
- Faba bean phyllody phytoplasma
- Peanut witches'-broom phytoplasma
- Sweet potato witches'-broom phytoplasma

16SrIII (X-disease group)
- Black locust witches'-broom phytoplasma
- Blueberry proliferation phytoplasma
- Canadian peach X phytoplasma
- Chayote witches'-broom phytoplasma
- Choke cherry X phytoplasma
- Cirsium white leaf phytoplasma
- Clover yellow edge phytoplasma
- Dandelion virescence phytoplasma
- Gaillardia phyllody phytoplasma
- Garlic decline phytoplasma
- Little peach phytoplasma
- Milkweed yellows phytoplasma
- Peach rosette phytoplasma
- Peach yellow leafroll phytoplasma
- Red suture phytoplasma
- Soybean veinal necrosis phytoplasma
- Strawberry leafy fruit phytoplasma
- Tsuwabuki witches'-broom phytoplasma
- Virginia grapevine yellows phytoplasma VGYIII
- Walnut witches'-broom phytoplasma
- Western X phytoplasma

16SrIV (Coconut lethal yellows group)
- Carludovica palmata leaf yellowing phytoplasma
- Coconut lethal yellowing phytoplasma
- Texas Phoenix palm phytoplasma
- Yucatan coconut lethal decline phytoplasma
- Phytoplasma sp. LfY1-Guerrero
- Phytoplasma sp. LfY2(Alv2)-Guerrero
- Phytoplasma sp. LfY4(MA3)-Oaxaca
- Phytoplasma sp. LfY4(MA5)-Oaxaca

Phytoplasma sp. LfY5(PE56)-Oaxaca
Phytoplasma sp. LfY5(PE65)-Oaxaca
16SrIX (Pigeon pea witches'-broom group)
Almond witches'-broom phytoplasma
Candidatus Phytoplasma phoenicium
Knautia arvensis phyllody phytoplasma
Lactuca serriola phytoplasma
Phaseolus lunatus phytoplasma
Pigeon pea witches'-broom phytoplasma
16SrV (Elm yellows group)
Alder yellows phytoplasma
Candidatus Phytoplasma ziziphi
Cherry lethal yellows phytoplasma
Elm yellows phytoplasma
Flavescence doree phytoplasma
Flavescence doree phytoplasma HD1
Hemp dogbane yellows phytoplasma
Jujube witches'-broom phytoplasma
Nectarine yellows phytoplasma
Peach yellows phytoplasma
Rubus stunt phytoplasma
Spartium witches'-broom phytoplasma
Virginia creeper phytoplasma
16SrVI (Clover proliferation group)
Candidatus Phytoplasma castaneae
Centaurea solstitialis virescence phytoplasma
Clover proliferation phytoplasma
Potato witches'-broom phytoplasma
16SrVII (Ash yellows group)
Candidatus Phytoplasma fraxini
Erigeron witches'-broom phytoplasma
16SrVIII (Loofah witches'-broom group)
Loofah witches'-broom phytoplasma
16SrX (Apple proliferation group)
Candidatus Phytoplasma mali
Candidatus Phytoplasma prunorum

Candidatus Phytoplasma pyri
Phytoplasma sp. 'ECA G32'
16SrXI (Rice yellow dwarf group)
Rice yellow dwarf phytoplasma
Sugarcane white leaf phytoplasma
16SrXII (Stolbur group)
Buckland Valley grapevine yellows phytoplasma
Candidatus Phytoplasma australiense
Lavender decline stolbur phytoplasma
Papaya dieback phytoplasma
Phormium yellow leaf phytoplasma
Stolbur P3L phytoplasma
Stolbur phytoplasma
Stolbur SBR phytoplasma
VK grapevine yellows phytoplasma
16SrXIII (Mexican periwinkle virescence group)
Chinaberry yellows phytoplasma
Mexican periwinkle virescence phytoplasma
16SrXIV (Bermuda white leaf group)
Bermuda grass white leaf phytoplasma
Chinese Bermudagrass white leaf phytoplasma
Date palm slow decline phytoplasma
White tip die-back phytoplasma
16SrXV (Hibiscus witches'-broom group)
Candidatus Phytoplasma brasiliense

Unclassified phytoplasmas

Allocasuarina muelleriana' phytoplasma
Fragaria multicipita' phytoplasma
Helichrysum bracteatum' phytoplasma
Momordica charantia' phytoplasma
Rehmannia glutinosa var. purpurea' phytoplasma
Vinca minor' phytoplasma
Alfalfa witches'-broom phytoplasma
Australian lucerne yellows phytoplasma
Beet leafhopper transmitted virescence phytoplasma
Bindweed yellows phytoplasma

Black raspberry witches'-broom phytoplasma
Brinjal little leaf phytoplasma
Cactus phytoplasma 'Martinez-Soriano 1999'
Cactus phytoplasma 'Martinez-Soriano 2001'
Cactus witches'-broom phytoplasma
Candidatus Phytoplasma rhamni
Carrot proliferation phytoplasma
Cenchrus bunchy shoot phytoplasma
Chinese pigeon pea witches'-broom phytoplasma
Chlorante-Aster yellows phytoplasma
Cocky apple witches'-broom phytoplasma
Cosmos bippinatus phyllody phytoplasma
Cuban Papaya phytoplasma
Cynodon white leaf phytoplasma
Elm phytoplasma 'Arlington Heights'
Elm witches'-broom phytoplasma
Ethiopian Gliricidia little leaf phytoplasma
Florida Crotolaria witches'-broom phytoplasma
Florida Pandanus decline phytoplasma
Florida periwinkle virescence phytoplasma
Florida periwinkle witches'-broom phytoplasma
Florida pigeon pea witches'-broom phytoplasma
Florida Rhynchosia little leaf phytoplasma
Honduran Gliricidia little leaf phytoplasma
Italian alfalfa witches-broom phytoplasma
Kerala wilt disease phytoplasma
Mexican coconut lethal decline phytoplasma
Oenothera phytoplasma
Oenothera phytoplasma 86-7
Omani alfalfa witches'-broom phytoplasma
Orange little leaf phytoplasma
Palm lethal yellowing phytoplasma
Periwinkle little leaf phytoplasma
Petunia dwarf phytoplasma
Picris echioides phyllody phytoplasma
Picris echioides yellows phytoplasma

Pigeon pea little leaf phytoplasma
Poinsettia branch-inducing phytoplasma
Protea spp. witches'-broom phytoplasma
Prunus armeniaca phytoplasma
Sorghum bunchy shoot phytoplasma
Sorghum grassy shoot phytoplasma
Sorghum grassy shoot phytoplasma variant I
Sorghum grassy shoot phytoplasma variant II
Spiraea stunt phytoplasma
Strawberry green petal phytoplasma
Strawberry lethal yellows phytoplasma
Stylosanthes little leaf phytoplasma
Sugarcane phytoplasma (uncharacterised)
Sugarcane phytoplasma type A
Sugarcane phytoplasma type B
Sugarcane phytoplasma type C
Sugarcane phytoplasma type D
Sugarcane phytoplasma type E
Sugarcane phytoplasma type F
Sugarcane phytoplasma type G
Sugarcane phytoplasma type H
Sugarcane phytoplasma type I
Sugarcane phytoplasma type J
Sugarcane Ramu stunt disease phytoplasma
Sugarcane yellows phytoplasma type I
Sunn-hemp phyllody phytoplasma
Sunshine tree witches'-broom phytoplasma
Sweet potato little leaf phytoplasma
Vigna little leaf phytoplasma
Weeping tea tree witches'-broom phytoplasma

Phytoplasma sp.

Phytoplasma sp. CWB
Phytoplasma sp. Gph
Phytoplasma sp. JWB-F
Phytoplasma sp. JWB-G1
Phytoplasma sp. JWB-G2

Phytoplasma sp. JWB-Ko
Phytoplasma sp. JWB-Ky
Phytoplasma sp. pathovar mosaic-inducing
Phytoplasma sp. PinP
Phytoplasma sp. PY1
Phytoplasma sp. PYCcwb
Phytoplasma sp. SPLL-V4
Phytoplasma sp. STRAWB1
Phytoplasma sp. STRAWB2
Phytoplasma sp. WDWB-K

Order—Anaeroplasmatales

Family—Anaeroplasmataceae

Genus—Anaeroplasma

Anaeroplasma abactoclasticum
Anaeroplasma bactoclasticum
Anaeroplasma varium

Genus—Asteroleplasma

Asteroleplasma anaerobium

Genera incertae sedis

Erysipelotrichaceae

Bulleidia

Bulleidia extructa
Bulleidia moorei

Erysipelothrix

Erysipelothrix inopinata
Erysipelothrix rhusiopathiae
Erysipelothrix tonsillarum
Erysipelothrix sp.

Holdemania

Holdemania filiformis

Solobacterium

Environmental samples

Order—Entomoplasmatales

Family—Entomoplasmataceae

Genus—Entomoplasma

Entomoplasma ellychniae
Entomoplasma freundtii

Entomoplasma lucivorax
Entomoplasma luminosum
Entomoplasma melaleucae
Entomoplasma somnilux

Genus—Mesoplasma

Mesoplasma chauliocola
Mesoplasma corruscae
Mesoplasma entomophilum
Mesoplasma florum
Mesoplasma grammopterae
Mesoplasma lactucae
Mesoplasma photuris
Mesoplasma pleciae
Mesoplasma seiffertii
cf. Mesoplasma symbiont of Acromyrmex octospinosus
Mesoplasma syrphidae
Mesoplasma tabanidae

Spiroplasmataceae

Spiroplasma

Spiroplasma apis
Spiroplasma citri
Spiroplasma clarkii
Spiroplasma diabroticae
Spiroplasma gladiatoris
Spiroplasma ixodetis
Spiroplasma kunkelii
Spiroplasma melliferum
Spiroplasma mirum
Spiroplasma monobiae
Spiroplasma phoeniceum
Spiroplasma poulsonii
Spiroplasma taiwanense
Spiroplasma sp.
Spiroplasma sp. MD-2
Spiroplasma sp. NI

Spiroplasma sp. NSRO
Spiroplasma sp. NSRO-A
Spiroplasma sp. SM
Spiroplasma sp. YR-1
Spiroplasma sp. YR-2

Unclassified Spiroplasmataceae

Spiroplasma-symbiont of Antonina crawii

Environmental samples

Uncultured Entomoplasmatales bacterium

Order—Mycoplasmatales

Family—Mycoplasmataceae

Genus—Mycoplasma

Candidatus Mycoplasma haemodidelphidis
Candidatus Mycoplasma haemolamae
Candidatus Mycoplasma haemominutum
Candidatus Mycoplasma haemoparvum
Candidatus Mycoplasma kahanei
Candidatus Mycoplasma ravipulmonis
Eperythrozoon ovis
Mycoplasma adleri
Mycoplasma agalactiae
Mycoplasma agassizii
Mycoplasma alkalescens
Mycoplasma alligatoris
Mycoplasma alvi
Mycoplasma anatis
Mycoplasma anseris
Mycoplasma arginini
Mycoplasma arthritidis
Mycoplasma auris
Mycoplasma BHJA
Mycoplasma bovigenitalium
Mycoplasma bovirhinis
Mycoplasma bovis
Mycoplasma bovoculi
Mycoplasma buccale

Mycoplasma buteonis
Mycoplasma californicum
Mycoplasma canadense
Mycoplasma canis
Mycoplasma capricolum
Mycoplasma caviae
Mycoplasma cavipharyngis
Mycoplasma citelli
Mycoplasma cloacale
Mycoplasma coccoides
Mycoplasma collis
Mycoplasma columbinasale
Mycoplasma columbinum
Mycoplasma columborale
Mycoplasma conjunctivae
Mycoplasma corogypsi
Mycoplasma cottewii
Mycoplasma cricetuli
Mycoplasma crocodyli
Mycoplasma cynos
Mycoplasma dispar
Mycoplasma edwardii
Mycoplasma elephantis
Mycoplasma equigenitalium
Mycoplasma equirhinis
Mycoplasma falconis
Mycoplasma fastidiosum
Mycoplasma faucium
Mycoplasma felifaucium
Mycoplasma feliminutum
Mycoplasma felis
Mycoplasma fermentans
Mycoplasma flocculare
Mycoplasma gallinaceum
Mycoplasma gallinarum
Mycoplasma gallisepticum

Mycoplasma gallopavonis
Mycoplasma gateae
Mycoplasma genitalium
Mycoplasma glycophilum
Mycoplasma gypis
Mycoplasma gypsbengalensis
Mycoplasma haemocanis
Mycoplasma haemofelis
Mycoplasma haemomuris
Mycoplasma hominis
Mycoplasma hyopharyngis
Mycoplasma hyopneumoniae
Mycoplasma hyorhinis
Mycoplasma hyosynoviae
Mycoplasma imitans
Mycoplasma indiense
Mycoplasma iners
Mycoplasma iowae
Mycoplasma lagogenitalium
Mycoplasma leonicaptivi
Mycoplasma leopharyngis
Mycoplasma lipofaciens
Mycoplasma lipophilum
Mycoplasma maculosum
Mycoplasma meleagridis
Mycoplasma microti
Mycoplasma moatsii
Mycoplasma mobile
Mycoplasma molare
Mycoplasma monodon
Mycoplasma muris
Mycoplasma mustelae
Mycoplasma mycoides
Mycoplasma neurolyticum
Mycoplasma opalescens
Mycoplasma orale

Mycoplasma ovipneumoniae
Mycoplasma oxoniensis
Mycoplasma penetrans
Mycoplasma phocicerebrale
Mycoplasma phocidae
Mycoplasma phocirhinis
Mycoplasma pirum
Mycoplasma pneumoniae
Mycoplasma pneumophila
Mycoplasma primatum
Mycoplasma pullorum
Mycoplasma pulmonis
Mycoplasma putrefaciens
Mycoplasma salivarium
Mycoplasma simbae
Mycoplasma spermatophilum
Mycoplasma spumans
Mycoplasma sturni
Mycoplasma sualvi
Mycoplasma subdolum
Mycoplasma suis
Mycoplasma synoviae
Mycoplasma testudineum
Mycoplasma testudinis
Mycoplasma timone
Mycoplasma verecundum
Mycoplasma wenyonii
Mycoplasma yeatsii
Mycoplasma zalophus
Mycoplasma sp.
Mycoplasma sp. 'bovine group 7'
Mycoplasma sp. 'ovine/caprine serogroup 11'
Mycoplasma sp. 94630
Mycoplasma sp. HRC689
Mycoplasma sp. PG50
Mycoplasma sp. serogroup L

Mycoplasma sp. VJC358
Environmental samples

Genus—Ureaplasma

Ureaplasma canigenitalium
Ureaplasma cati
Ureaplasma diversum
Ureaplasma felinum
Ureaplasma gallorale
Ureaplasma loridis
Ureaplasma parvum
Ureaplasma urealyticum.

3

MYCOPLASMA

Mycoplasmas are a specific and unique species of bacteria - the smallest free-living organism known on the planet. The primary differences between mycoplasmas and other bacteria is that bacteria have a solid cell-wall structure and they can grow in the simplest culture media. Mycoplasmas however, do not have a cell wall, and like a tiny jellyfish with a pliable membrane, can take on many different shapes, which make them difficult to identify, even under a high-powered electron microscope. Mycoplasmas can also be very hard to culture in the laboratory and are often missed as pathogenic causes of diseases for this reason. It is strictly anaerobic and has a special characteristic of metabolizing glucose. Like the virus, it is very dependent upon its host for survival. It was originally classified as a virus due to its small size.

Over the past decades cell culturing has become an indispensable tool for modern research. With the increased use of cell cultures contaminations are increasingly frequent, i.e. cellular contamination (cross-contamination between cell cultures) and contamination with microorganisms (microbial contamination). While bacterial and fungal infections of cultures are relatively easy to detect, to prevent and to treat, contamination with mycoplasmas represents a much bigger problem in terms of incidence, detectability, prevention, eradication and effects. It was estimated that between 5 and 35% of cell cultures in current use are infected with mycoplasmas. Mycoplasma contamination is not without serious consequences as virtually every cellular process can be altered by the infection. Mycoplasmas compete effectively with tissue-culture cells for medium nutrients, thus depriving the cells of essential nutrients, resulting in profound effects on cell metabolism and function. In the worst scenario, contamination leads to diminished cell growth and eventually to the loss of cultures. Mycoplasmas are wall-less bacteria belonging to various genera within the class Mollicutes. These flexible, pleomorphic organisms can be as small as 0.2-0.3μm and can achieve very high densities in cell cultures, (107 - 108 organisms/mL), without discernable pH changes or turbidity. Mycoplasmas can usually be grown on artificial media, but most species require very complex, nutritionally enriched media and carefully defined environmental conditions. There are currently more than 120 species in 6 genera, many of which are pathogenic. The vast majority of cell culture contaminants belong to only 6 species of human, bovine or porcine origin.

These very common contaminants are: *Mycoplasma hyorhinis*, *Mycoplasma arginini*, *Mycoplasma salivarium*, *Mycoplasma orale*, *Mycoplasma fermentans* and *Acholephasma laidlawii*. Mycoplasmas from human, bovine and porcine sources are the most prevalent groups with *Acholeplasma laidlawii*, *Mycoplasma arginini*, *Mycoplasma fermentans*, *Mycoplasma hyorhinis*, and *Mycoplasma orale* being the single most common isolates. Numerous methods for detecting mycoplasma infection have been described. While some tests are rather cumbersome and time-consuming, other less fastidious techniques do not have sufficient sensitivinty and specificity. Mycoplasma have a cellular-like structure (elementary body) without a true cell wall. They contain protein, DNA, RNA, and enzymes. The elementary bodies vary in shape and size; some are 1/2,500 of an inch long and others are 1/10 that size. Also, mycoplasma are sensitive to heat and some antibiotics. Mycoplasmas are the smallest free-living, self-replicating organisms (0.2-2 μm in diameter). Mycoplasmas lack a cell wall so that antibiotics interfering with the murein formation of cell walls (such as penicillin which is often added to cell cultures) are not effective at the standard concentrations used. Mycoplasmas in cell cultures are extracellular parasites usually attached to the external surface of the cell membrane.

Surveys and research studies have repeatedly shown that 15-50% or more of all cell lines are contaminated with one or more mycoplasmal species, often at concentrations of 107 - 108 colony forming units per milliliter of culture fluid. Contamination rates are even higher for cell lines routinely grown in antibiotic-containing medium.

Mycoplasmas have been shown to alter virtually every property and parameter measured in cell cultures, including hybridoma selection rates, protein and nucleic acid synthesis, immunogenicity, chromosomal breakage and production of virus and biologicals such as interferon and monoclonal antibodies. As a result of this widespread problem, bio production and research is often unknowingly done using mycoplasmal contaminated cell cultures. The validity and significance of research and the safety of the biologicals produced from contaminated cell cultures must be questioned.

The high mycoplasmal contamination rate of cell cultures is primarily due to four factors:

1. The absence of visible signs of mycoplasmal contamination including changes in turbidity, cytopathic effect and pH, even in heavily contaminated cultures, leads to a false sense of security. Once present in the laboratory, mycoplasmally infected cultures often cross-contaminate the other cell lines being used.
2. The continuous use of antibiotics, which are not usually mycoplasmacidal in long-term cell cultures, can mask poor aseptic technique. This leads to accidental introduction of mycoplasmas along with other microorganisms.
3. The lack of simple, easy-to-use and reliable detection methods in the past has resulted in many cell lines being untested. Because of this, many cell lines brought into a lab from outside sources although appearing normal and healthy, are in fact already infected with mycoplasma.
4. Because of their small size and lack of cell wall, mycoplasmas occasionally penetrate the filters used to "sterilize" cell culture media and sera, resulting in low levels of these organisms being accidentally introduced into cultures during routine feeding.

The accepted name was chosen because Mycoplasmas were observed to have a fungi-like structure (Mycology is the study of fungi – hence "Myco") and it also had a flowing plasma-like structure without a cell wall – hence "plasma". The first strains were isolated from cattle with arthritis and pleuro-pneumonia in 1898 at the Pasteur Institute. The first human strain was isolated in 1932 from an abscessed wound. Drs. Swift and Brown made the first connection between mycoplasmas and rheumatoid diseases in 1939. Unfortunately, mycoplasmas didn't become part of the medical school curriculum until the late 1950s when one specific strain was identified and proven to be the cause of atypical pneumonia, and named Mycoplasma pneumonia. The association between immunodeficiency and autoimmune disorders with mycoplasmas was first reported in the mid 1970s in patients with primary hypogammaglobulinemia (an autoimmune disease) and infection with four species of mycoplasma that had localized in joint tissue. Since that time, scientific testing methodologies have made critical technological progress and along with it, more mycoplasma species have been identified and recorded in animals, humans and even plants.

Mycoplasmas are considered to belong to a distinct order (Mycoplasmatales) in the class Mollicutes, phylum Firmicutes of the Bacteria. They are very small (mostly 0·1 to 1 μm), have flexible cells bounded by a 3-layered membrane, not a true cell wall, and cannot secrete cell wall precursors like muramic acid. They are highly variable in shape (pleomorphic), coccoid to filamentous; usually non-motile; saprophytic and parasitic. Reproduction is controversial, but there seem to develop within filaments tiny coccoid structures called elementary bodies, released by fragmentation of filaments and/or binary fission; budding also occurs. Mycoplasmas are normally associated with animals, and they occur extracellularly in vertebrates.

The balance of the molecular evidence, based largely on comparison of base sequences of the highly conserved ribosomal RNA (rRNA) molecules, particularly of the 16S rRNA type, favors the hypothesis of degenerative evolution. According to Woese and his colleagues, the mycoplasmas evolved as a branch of the low-guanine-plus-cytosine Gram-positive bacteria and are most closely related to two clostridia, Clostridium innocuum and C ramosum. However, the marked phenotypic and genotypic variability among mycoplasmas has led some workers to conclude that myco-plasmas evolved from a variety of walled bacteria and accordingly have a polyphyletic origin. Woese maintains that the origin of mycoplasmas is monophyletic and explains the great variety of mycoplasmas by a process of rapid evolution characteristic of the group.

General Characteristics

The mycoplasmas are essentially bacteria lacking a rigid cell wall during their entire life cycle, although they are also much smaller than bacteria. The first organism of this type was associated with pleuropneumonia of cattle, and was originally called the pleuropneumonia organism (PPO). Since that time, a number of organisms with similar morphological characteristics and cultural properties have been isolated. These are commonly referred to as pleuropneumonia-like organisms or PPLO. A certain group of mycoplasmas produce extremely tiny colonies on agar plates, and are called the T-strains.

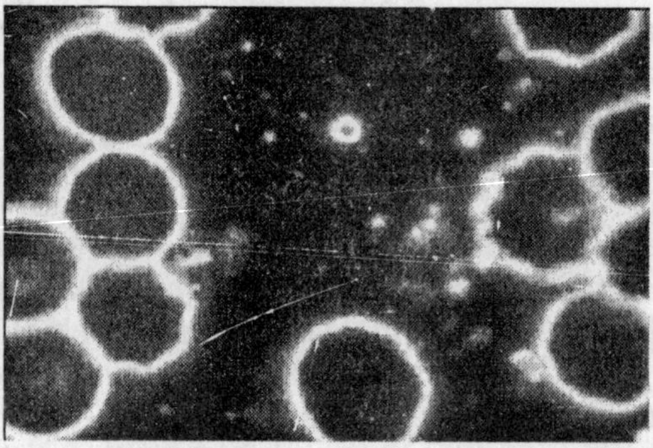

Fig. 3.1. Mycoplasma seen using darkfield microscopy.

Some bacteria readily give rise spontaneously to variants that can replicate in the form of small filterable protoplasmic elements with defective or absent cell walls. These organisms, called L-forms, can also be formed by many species when cell wall synthesis is impaired by antibiotic treatment or high salt concentration. These organisms differ from mycoplasma in that they contain a rigid cell wall, at least at one stage of their life cycle and contain no sterols in their cytoplasmic membrane.

These organisms are the smallest known free-living organisms. Because of the absence of cell walls, they do not stain with the Gram stain, and they are more pleomorphic and plastic than eubacteria. With Giemsa stain, they appear as tiny pleomorphic cocci, short rods, short spirals, and sometimes as hollow ring forms. Their diameter ranges from 0.15 μm to 0.30 μm.

Most mycoplasmas require a rich medium containing a sterol and serum proteins for growth. Despite the lack of a cell wall, they do not require a medium of very high osmotic pressure. On solid media, they form minute, transparent colonies. When viewed under low-power magnification, the colony looks like a fried egg. The different strains vary in their growth rate and may take from two days to several weeks to form a colony. The mycoplasmas are the smallest free-living bacteria. They range from 0.2 - 0.8 micrometers and thus can pass through some filters used to remove bacteria. They have the smallest genome size and, as a result, lack many metabolic pathways and require complex media for their isolation. The mycoplasmas are facultative anaerobes, except for *M. pneumoniae*, which is a strict aerobe. The mycoplasmas grow slowly by binary fission and produce "fried egg" colonies on agar plates; the colonies of *M. pneumoniae* have a granular appearance. Due to the slow growth of mycoplasmas, the colonies may take up to 3 weeks to develop and are usually very small. The colonies of Ureaplasma are extremely small and thus *Ureaplasma* are also called T-strains (tiny strains). The mycoplasmas all require sterols for growth and for membrane synthesis. The three species can be differentiated by their ability to metabolize glucose (*M. pneumoniae*), arginine (*M. hominis*) or urea (*U. urealyticum*). The fourth species *M. genitalium* is extremely difficult to culture.

Although there are many species of mycoplasmasas described in Chapter 2 in this book, only four are recognized as human pathogens; *Mycoplasma pneumoniae*, *Mycoplasma hominis*, *Mycoplasma genitalium*, and *Ureaplasma urealyticum*. Although there are other species that have been isolated from humans, their role in disease is not well established.

Table 3.1. The diseases caused by *M. pneumoniae, M. hominis, M. genitalium* and *U. urealyticum*.

Organism	*Disease*
M. pneumoniae	Upper respiratory tract disease, tracheobronchitis, atypical pneumonia
M. hominis	Pyelonephritis, pelvic inflammatory disease, postpartum fever
M. genitalium	Nongonococcl urethritis
U. urealyticum	Nongonococcl urethritis

Structure

Morphologically, the Mycoplasma appear as very small cocci (300 nm diameter). There are also elongated and filamentous forms. The mycoplasma cell contains the minimum set of organelles essential for growth and replication: a plasma membrane, ribosomes, and a genome consisting of a double-stranded circular DNA molecule without nuclear membrane but lacks mesosomes.

A limiting membrane encloses the cell, which is more similar to that of animal cells than that of bacterial cells because of sterols present in the membrane. In some strains, amorphous material on the outer surface of the membrane suggests the existence of a capsule.

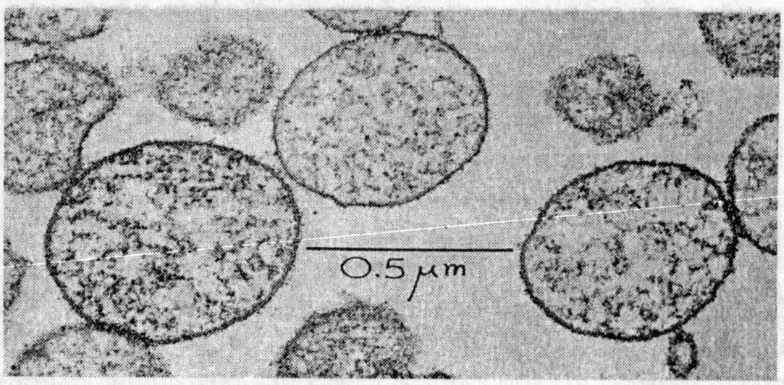

Fig. 3.2. Electron micrograph of thin-sectioned mycoplasma cells.

Mycoplasmas have been nicknamed the "crabgrass" of cell cultures because their infections are persistent, frequently difficult to detect and diagnose, and difficult to cure. Contamination of cell cultures by mycoplasmas presents serious problems in research laboratories and in biotechnological industries using cell cultures. The origin of contaminating mycoplasmas is in components of the culture medium, particularly serum, or in the flora of the technician's mouth, spread by droplet infection.

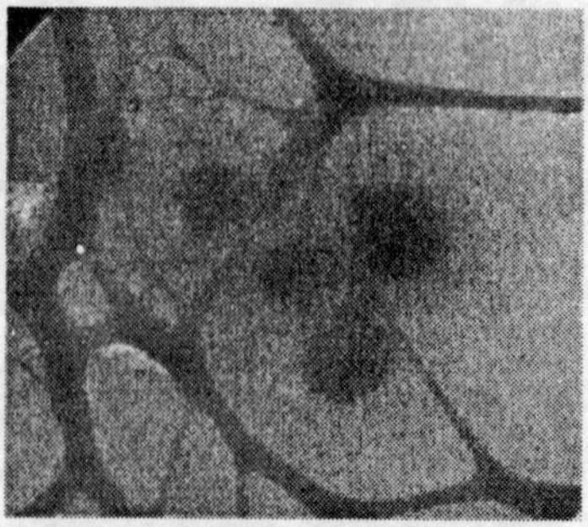

Fig. 3.3. Frozen cell.

Some mycoplasmas possess unique attachment organelles, which are shaped as a tapered tip

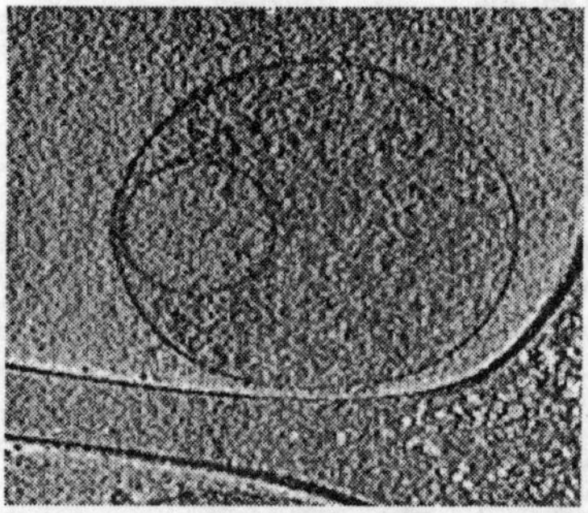

Fig. 3.4. Slice of reconstruction.

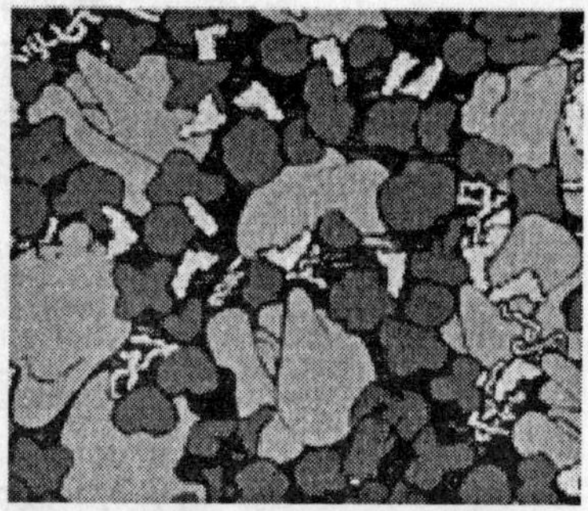

Fig. 3.5. Crowdedness schematic.

in *M. pneumoniae* and *M genitalium*. *Mycoplasma pneumoniae* is a pathogen of the respiratory tract, adhering to the respiratory epithelium, primarily through the attachment organelle. Interestingly, these two human mycoplasmas exhibit gliding motility on liquid-covered surfaces. The tip structure always leads, again indicating its importance in attachment.One of the most useful distinguishing features of mycoplasmas is their peculiar fried-egg colony shape, consisting of a central zone of growth embedded in the agar and a peripheral one on the agar surface.

The organisms possess a protein attachment factor termed P1 that interacts with a specific cellular receptor and allows adherence to respiratory epithelium. On agar plates, *Mycoplasma* have a "fried egg" colonial morphology. The organisms have limited biosynthetic abilities; they require cholesterol for their cell membrane and can generate energy via the breakdown of arginine. *Ureaplasma* requires urea to produce an electrochemical gradient; urea is converted to ammonia to produce ATP. These bacteria have the smallest genome of any prokaryote (about 20% that of *E. coli*) and the lowest GC content (about 24%). The lack of cell walls and intracytoplasmic membranes facilitates isolation of the mycoplasma membrane in a relatively pure form. The isolated mycoplasma membrane resembles that of other prokaryotes in being composed of approximately two-thirds protein and one-third lipid. The mycoplasma lipids resemble those of other bacteria, apart from the large quantities of cholesterol in the sterol-requiring mycoplasmas.

Membrane proteins, glycolipids, and lipoglycans exposed on the cell surface are the major antigenic determinants in

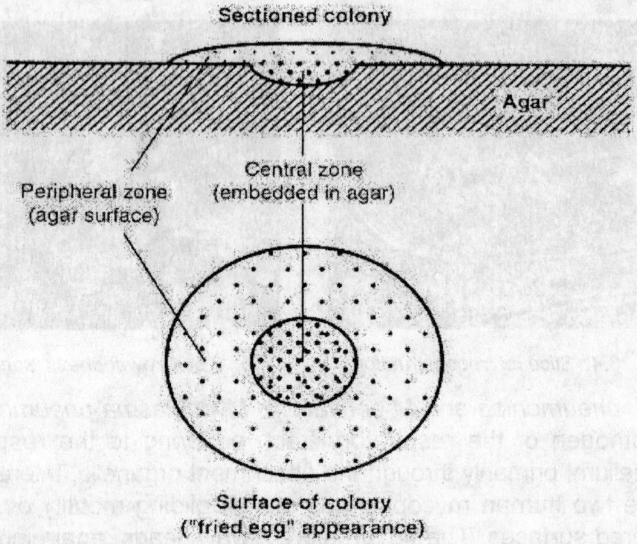

Fig. 3.6. Morphology of a typical "fried-egg" mycoplasma colony.

mycoplasmas. Antisera containing antibodies to these components inhibit growth and metabolism of the mycoplasmas and, in the presence of complement, cause lysis of the organisms. These properties are used in various serologic tests that differentiate between mycoplasma species and serotypes and detect antibodies to mycoplasmas in sera of patients

Some mycoplasmas have the distinction of being the smallest living organisms. The DNA sequences of the complete genomes of three mycoplasmas have been determined:

- *Mycoplasma genitalium* has 580,073 base pairs of DNA encoding 517 genes (480 for proteins; the rest for RNAs).
- *Mycoplasma urealyticum* has 751, 719 base pairs of DNA encoding 651 genes (613 for proteins; 39 for RNAs).
- *Mycoplasma pneumoniae* has 816,394 base pairs of DNA encoding 679 genes.

The mycoplasma genome is typically prokaryotic, consisting of a circular, double stranded DNA molecule. The *Mycoplasma* and *Ureaplasma* genomes are the smallest recorded for any self-reproducing prokaryote. Therefore, there are very few genes; in some mycoplasmas the number is estimated at fewer than 500, about one sixth the number of genes in *Escherichia coli*.

Mycoplasmas accordingly express a small number of cell proteins and lack many enzymatic activities and metabolic pathways. Their nutritional requirements are correspondingly complex, and they are dependent on a parasitic mode of life.

The dependence of mycoplasmas on their host for many nutrients explains the great difficulty of cultivation in the laboratory. The complex media for mycoplasma culture contain serum, which provides fatty acids and cholesterol for mycoplasma membrane synthesis. The requirement of most mycoplasmas for cholesterol is unique among prokaryotes. The consensus is that only a small fraction of mycoplasmas existing in nature have been cultivated so far. Some of the cultivable mycoplasmas, including the human pathogen *M. pneumoniae*, grow very slowly, particularly on primary isolation. *Ureaplasma urealyticum*, a pathogen of the human urogenital tract, grows very poorly in vitro, reaching maximal titers of 107 organisms/ml of culture. *Mycoplasma genitalium*, another human pathogen, grows so poorly in vitro that only a few successful isolations have been achieved.

Glucose and other metabolizable carbohydrates can be used as energy sources by the fermentative mycoplasmas possessing the Embden-Meyerhof-Parnas glycolytic pathway. All mycoplasmas examined thus far possess a truncated, flavin-terminated respiratory system, which rules out oxidative phosphorylation as an ATP-generating mechanism. Breakdown of arginine by the arginine dihydrolase pathway has been proposed as a major source of ATP in nonfermentative mycoplasmas. Ureaplasmas have a requirement, unique among living organisms, for urea. Because they are non-glycolytic and lack the arginine dihydrolase pathway, it has been suggested, and later proven experimentally, that ATP is generated through an electrochemical gradient produced by ammonia liberated during the intracellular hydrolysis of urea by the organism's urease.

The mycoplasma genome is characterized by a low guanine-plus-cytosine content and by a corresponding preferential utilization of codons containing adenine and uracil, particularly in the third position. Most interesting is the use of the universal stop codon UGA as a tryptophan codon in many mycoplasmas, a rare property found so far only in mycoplasmas and in nonplant mitochondria. Resistance of mycoplasmal RNA polymerase to rifampicin is another property distinguishing mycoplasmas from the conventional eubacteria. However, apart from this resistance to rifampicin, the mycoplasmas are susceptible to antibiotics, such as tetracyclines and chloramphenicol, that inhibit protein synthesis on prokaryotic ribosomes.

Mycoplasmal pneumonia

The term primary atypical pneumonia was coined in the early 1940s to describe pneumonias different from the typical lobar pneumonia caused by pneumococci. Several common respiratory viruses, including influenza virus and adenovirus, were shown to be responsible for a significant number of these pneumonias. From other cases, many of which developed antibodies agglutinating red blood cells in the cold (cold agglutinins), an unidentified filterable agent was isolated by Eaton and associates and was called Eaton agent. This agent was identified as a new Mycoplasma species after its successful cultivation on cell-free media in 1962. Named *Mycoplasma pneumoniae*, it was the first clearly documented mycoplasma pathogenic for humans.

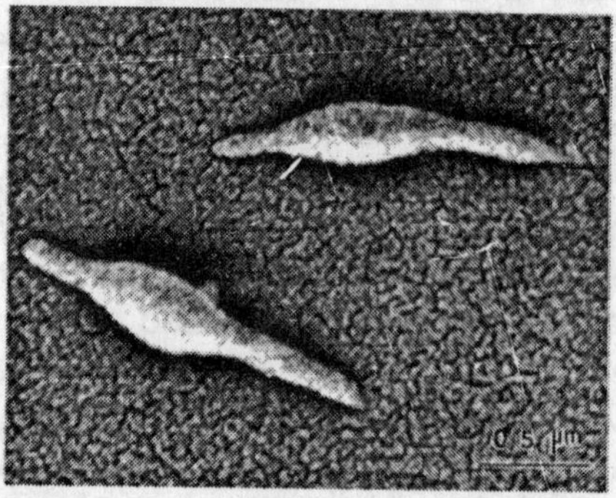

Fig. 3.7. Morphology of a typical "fried-egg" mycoplasma colony.

The effects of *M. pneumoniae* on humans include subclinical infection, upper respiratory disease, and bronchopneumonia. Most human infections do not progress to a clinically evident pneumonia. When pneumonia occurs, the onset generally is gradual and the clinical picture is one of a mild to moderately severe illness, with early complaints referable to the lower respiratory passages. Radiography frequently reveals evidence of pneumonia before physical signs are apparent. Involvement is usually limited to one of the lower lobes of the lungs, and the pneumonia is interstitial or bronchopneumonic. The course of disease varies; remittent fever, cough, and headache persist for several weeks. One of the most consistent clinical features is a long convalescence, which may extend from 4 to 6 weeks. Few fatal cases have been reported. Several unusual complications have been noted, including hemolytic anemia, polyradiculitis, encephalitis, aseptic meningitis, and central nervous system illness such as Guillain-Barré syndrome. In addition, pericarditis and pancreatitis have been observed. These sequelae may be related to the suspected immunopathology of M pneumoniae disease.

The complete sequence of the *M. pneumoniae* genome (Himmelreich et al., 1996) shows that it is a circular duplex DNA of 816,394 bp. The presumptive origin of replication is in an A+T rich region between dnaA and dnaN. A total of 677 open reading frames (ORFs) and 39 genes coding for various RNAs were identified. Only 67 (or 9.9% of the total) ORFs had no significant similarity to sequences in the databases. The predicted coding regions that could be compared with both *Eschericia coli* and *Bacillus subtilis* showed higher similarity to the *B. subtilis* sequence, strongly supporting the deduced evolutionary relationship between *Mycoplasma* and the *Lactobacillus-Clostridium* branch of the gram-positive eubacteria.

Similar to the case for *M. genitalium*, one would expect the evolution of *M. pneumoniae* as a parasitic bacterium to allow the loss of biosynthetic pathways that were no longer needed, while retaining the transport proteins needed for acquisition of essential building blocks for macromolecular synthesis (e.g., amino acids, purines, pyrimidines, fatty acids) from the host. Indeed, the reductive evolution of *M. pneumoniae* from ancestral bacteria resulted in a smaller genome. The main causes of this reduction in genome size are (1) the loss of several anabolic pathways, such as biosynthesis of amino acids, purines and pyrimidines, and fatty acids, (2) the absence of a capacity to make a cell wall, and (3) a reduction of the number of proteins involved in essential processes such as DNA replication, repair, recombination, cell division, and protein secretion. A total of 44 transport proteins have been predicted from the genome sequence. This modest number, compared to the number of substrates expected to be transported, suggests that many of these transporters may not be highly specific.

The coding regions in *M. pneumoniae* comprise a total length of 724,174 bp, or 88.7% of the genome. The average size of a gene is 1011 bp, giving an average of one gene every 1140 bp. This is similar to the gene size and density seen with both the smaller M. genitalium genome as well as the H. influenza genome, which is more than twice as large.

Mycoplasma genitalium

The complete sequence of the *M. genitalium* genome (Fraser et al., 1995) shows that it is a circular duplex DNA of 580,070 bp. The presumptive origin of replication is in an A+T rich region between dnaA and dnaN. A total of 470 open reading frames (ORFs) were identified, with about half the genome transcribed predominantly in one direction from the origin and the other half transcribed predominantly in the opposite direction. Thus this genome shows a symmetrical strand bias for transcription reminiscent of some circular viral genomes. Of the 470 ORFs, 374 were identified by sequence matches to entries in a non-redundant bacterial protein database or in the complete set of translated sequences from the Haemophilus influenza genome, or by GeneMark. 96 ORFs have no sequence matches in the databases. The predicted coding regions that could be compared with both *Eschericia coli* and *Bacillus subtilis* showed higher similarity to the *B. subtilis* sequence, strongly supporting the deduced evolutionary relationship between *Mycoplasma* and the *Lactobacillus-Clostridium* branch of the gram-positive eubacteria.

The 470 ORFs in *M. genitalium* show an average size of 1040 bp and comprise 88% of the genome, giving an average of one gene every 1235 bp. This is similar to the gene size and density seen with the 1727 predicted coding regions in *H. influenza*, with an average gene size of 900 bp comprising 85% of the genome at a density of one gene every 1042 bp. Thus the reduction in genome size for *M. genitalium* did not result from a increase in gene density or reduction in gene size. One major factor in reducing genome size of this parasitic microbe is the substantial loss of genes encoding enzymes needed for many anabolic pathways, in particular biosynthesis of amino acids, purines and pyrimidines, and fatty acids. Also, the mycoplasma have lost the capacity to make a cell wall.

Not surprisingly for a parasitic organism that must acquire most of its cellular building blocks from its host, a substantial number of transport proteins are encoded, with a fraction of total genes in this category similar to that of H. influenza. As expected, genes encoding enzymes needed for replication and repair of DNA, transcription, translation and "cellular processes" such as cell division, cell killing, and protein secretion are present, again at a fraction of the total genes similar to (or greater than) that of *H. influenza*. Catabolic metabolism to produce energy is largely anaerobic, with an intact glycolytic pathway but not tricarboxylic acid cycle or electron transport system (Fraser et al., 1995).

It has been argued that the "degenerative" mode of evolution of *M. genitalium* by reduction in genome size from that of an ancestral gram-positive bacterium has led to the minimal set of genes needed for a living organism. One might expect that the "necessary" genes that are retained will provide a similar function and hence be conserved in other bacteria or other organisms. If so, then the removal of "unnecessary genes" will increase the fraction of proteins conserved in other species. The similar fraction of total genes devoted to replication, transcription and other retained processes in both *H. influenza* and *M. genitalium* (Fraser et al., 1995) shows that the complement of proteins devoted to these tasks could be reduced by about the same amount in *M. genitalium*. By searching for "bacterial conserved regions" (or BCRs) conserved in distantly related bacteria and for "ancient conserved regions" (or ACRs) shared with eukaryotic or archaeal homologs, Koonin et al. (1996) showed that the fraction of proteins containing BCRs and ACRs is close to the same in *E. coli*, *H. influenza*, and *M. genitalium*. This refutes the model that only a core of critical genes is retained in *M. genitalium*, but rather suggests that in all three bacteria there is a balance between highly conserved genes and more variable genes. This has been rationalized as possibly reflecting an equilibrium between the stability of major physiological processes and need for environmental adaptability (Koonin et al., 1996).

M. hominis is carried by ~15% of sexually active men and women. Usually these organisms behave as commensals, but they are sometimes associated with pelvic inflammatory disease, postpartum fever, pyelonephritis. *Ureaplasma urealyticum* (urease producer) - colonizes 60% of sexually-active women asymptomatically and the urethra of 5-20% of asymptomatic sexually active men. May cause some cases of nongonococcal urethritis.

However, *Ureaplasmas* have also been associated with pelvic inflammatory disease, as well as spontaneous abortions and low-birth-weight infants. Colonization of amniotic fluid with *U. urealyticum* is associated with meningitis and other CNS complications; introduction of the organism into the oropharynx at birth plays in important role in the development of neonatal pneumonia and other respiratory complications.

The so-called *Mycoplasma mycoides* cluster consists of six species or subspecies of mycoplasmas (Mollicutes). These species are pathogenic for ruminants and some of them are of great concern in veterinary medicine. The members of the *M. mycoides* cluster have two rRNA operons (rrnA and rrnB). The nucleotide sequences of the 16S rRNA genes of 10 strains, representing all of the known species and subspecies of the *M. mycoides* cluster, were determined by direct automated solid-phase DNA sequencing. The sequences of both rRNA operons were determined by a novel strategy involving in vitro amplification by PCR with one operon-specific primer pair and one general primer pair. Interestingly, sequence differences (polymorphisms) between the two operons were observed for all strains. Two strains of *M. capricolum* subsp. *capripneumoniae* were sequenced, and 15 polymorphisms were found in the type strain (F38) and 17 polymorphisms were found in the other strain (4/2LC). Eight polymorphisms were found in the 16S rRNA genes of the *M. mycoides* subsp. *mycoides* small-colony type, and sequence length variations in a poly(A) region were observed in the 16S rRNA genes of the two operons of this species. Secondary-structure analysis showed that polymorphisms were present in both stem and loop regions. The nucleotide substitutions in the polymorphic sites of the stem regions often resulted in a change from a canonical to a noncanonical base pairing or vice versa. A compensatory mutation was never observed in the other nucleotide of the base pair. Phylogenetic analysis based on the 16S rRNA sequences indicated that *Mycoplasma* sp. strain PG50 should be included in the *M. capricolum* species group. Furthermore, the 16S rRNA sequences of *M. mycoides* subsp. capri and the *M. mycoides* subsp. mycoides large-colony type were 99.9% identical.

Metabolism

Mycoplasma are flask-shaped and are most likely descended from Gram-positive bacteria. Due to their seriously degraded genome they cannot perform many metabolic functions, such as cell wall production or, synthesis of purines. As such stripped down organisms they are considered the perfect model of the minimalist cell. Meaning they are believed to contain the absolute minimum machinery necessary for survival and are considered the model organisms for the essential functions of all living cells. The mycoplasma cell is built of a minimum set of organelles including a plasma membrane, ribosomes, and a highly coiled circular chromosome.

Because it is essential for mycoplasma to be able to bind to their host cells they have developed special tip organelles for this purpose, with a significant percentage of genes in their genome devoted to this important function. Most species of mycoplasma are extracellular pathogens, and use their tip organelle,which has a high concentration of adhesins, to attach to the eukaryotic cell. The organelle is bounded by the cell membrane and contains a central rod-shaped structure. Mycoplasma that act as intracellular pathogens, penetrate their host cell using their tip organelle. The lack of a rigid cell wall, may also help facilitate contact between mycoplasma and their host cell, by creating the possibility of fusion between the two membranes. This would enable the exchange of membrane and cytoplasmic components.

The Mycoplasma represent the smallest and simplest self-replicating bacteria. They have no cell wall and are parasites that inhabit the mucus membranes of the respiratory and urogenital tracts. The parasitic mycoplasmas have truncated respiratory systems, lacking quinones and cytochromes. Another indication for the simplicity of the electron transport chain is the finding that the reduced nicotinamide adenine dinucleotide (NADH) oxidase activity is cytoplasmic. Complex electron transport chains are usually membrane bound, since they depend on the spatial organization of their components. Ruling out oxidative phosphorylation as an ATP-generating system leaves only two proven ways of ATP generation, both based on substrate level phosphorylation. The major source for ATP is the arginine dihydrolase pathway.

$$\text{Arginine} + H_2O \xrightarrow{\text{arginine deaminase}} \text{citrulline} + NH_3$$

$$\text{Citrulline + inorganic orthophosphate} \xrightarrow{\text{ornithine carbamoyltransferase}} \text{ornithine + carbamoylPO}$$

$$\text{Carbamoyl } PO_4 + ADP \xrightarrow{\text{carbamate kinase}} ATP + CO_2 + NH_3$$

Another mechanism for ATP generation is:

$$\text{Acetyl CoA + inorganic orthophosphate} \xrightarrow{\text{phosphate acetyltransferase}} \text{acetyl } PO_4 + \text{CoA}$$

$$\text{Acetyl PO4+ ADP} \xrightarrow{\text{acetate kinase}} \text{Acetate + ATP}$$

Acetyl CoA is produced by oxidative decarboxylation of pyruvate.

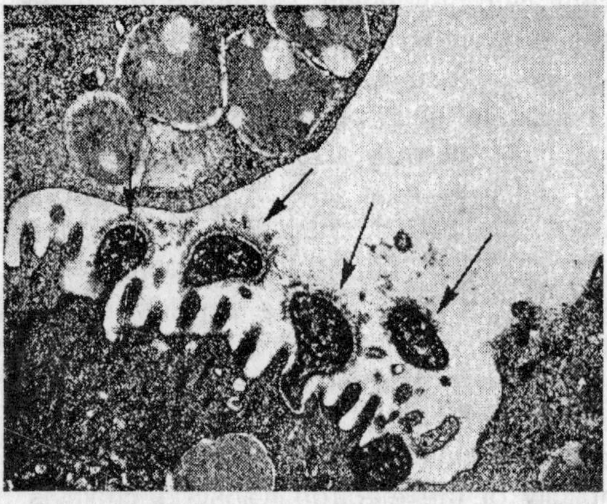

Fig. 3.8. Photomicrograph of the nasal cavity of a desert tortoise. Multiple mycoplasma are attached to the cell surface.

A few species derive their energy from the degradation of glucose or the hydrolysis of urea. All species synthesize DNA, RNA, lipids and proteins. However, it is not known if they can synthesize amino acids. Those species that require sterols incorporate these sterols (mainly cholesterol) into the cell membrane up to concentrations of 65%.

Cell Reproduction Cycle

The coccus is the basic form of all mycoplasmas in culture. The diameter of the smallest coccus capable of reproduction is about 300 nm. In most mycoplasma cultures, elongated or filamentous forms (up to 100 µm long and about 0.4 µm thick) also occur. The filaments tend to produce truly branched mycelioid structures, hence the name mycoplasma (myces, a fungus;

plasma, a form). Mycoplasmas reproduce by binary fission, but cytoplasmic division frequently may lag behind genome replication, resulting in formation of multinuclear filaments.

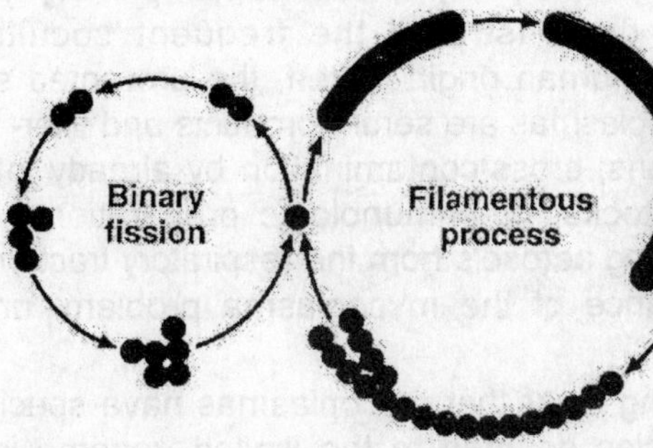

Fig. 3.9. Schematic presentation of the mode of mycoplasma reproduction.

In the absence of a rigid cell wall, the pattern of replication is quite different from that of typical bacteria, whose division starts with the formation of a well-defined septum. Though the mechanism of division in mycoplasmas is controversial, sequential microscopic observation suggests that new elementary particles arise by fragmentation of filamentous cells containing several discrete DNA components. Mycoplasmas are evolutionary descendants of the low G+C containing gram-positive bacteria and, through chromosome reduction, represent the smallest self-replicating life forms. Their streamlined genome size, which illustrates extreme biological gene economy, imposes complex nutritional requirements, such as dependence on external supplies of biosynthetic precursors, including amino acids, nucleotides, fatty acids, and sterols. This limited coding capacity dictates for mycoplasmas a parasitic way of life that few pathogenic microorganisms can claim. Therefore, the view that pathogenic mycoplasmas can grow “independently” requires an appreciation of their fastidious nature and their intimate dependence upon the host. Because of these properties, pathogenic mycoplasmas are among the most difficult micro-organisms to grow from clinical specimens and remain frequent contaminants of primary and continuous eucaryotic cell lines and tissue cultures. In some instances, mycoplasma contamination is obvious since infected eucaryotic cells exhibit aberrant growth, metabolism, and morphology. However, mycoplasmas often establish covert and chronic infections of target cells that lead to either invalid and misleading data or introduction of mycoplasmas or their products into reagents dedicated to therapeutic or research purposes. The recent emphasis on isolating viral agents, such as human immunodeficiency virus (HIV)-1, from human primary lymphocytic cells has also demonstrated the frequent cocultivation of mycoplasmas of human origin. Often, the unwanted sources of exogenous mycoplasmas are serum products and filter- "sterilized" (450 nm) solutions; cross-contamination by already infected cell cultures, viral stocks, or immunologic preparations; breaks in technique, including aerosols from the respiratory tract or by mouth pipetting; ignorance of the mycoplasma problem; or scientific indifference.

It is becoming clear that mycoplasmas have specialized cell reproduction cycles adapted to the limited genome information and parasitic life. Mycoplasmas are prokaryotes with the general characteristics and cellular organization of bacteria, but in contrast to bacteria they have no cell walls and they are the smallest known cells capable of independent replication. Mycoplasmas are parasitic bacteria that have extremely low G+C contents and small genomes. Their morphology is irregular because of the lack of a peptidoglycan layer. In *Escherichia coli*, initiation of chromosomal DNA replication occurs once during the cell's replicative cycle, and the nucleoids partition before cell division. The chromosornal replication of *E. coli* initiates in a small region and proceeds in both directions. It is mainly controlled by the timing and frequency of initiation, while the velocity of replication is constant. In mycoplasmas, chromosome replication also starts at a fixed site, followed by bidirectional progression. As in many eubacteria, the dnaA gene is expressed and plays important roles in the initiation of replication. These observations suggest that the outline of chromosome replication of mycoplasmas is similar to that of *E. coli*. However, the process of mycoplasma cell reproduction has not been clarified. Moreover, the cell division cycle of *E. coli* cannot be simply applied to mycoplasmas because of their irregular cell morphology. A model has been suggested for the cell cycle of *Mycoplasma mycoides*, which is closely related to *Mycoplasma capricolum*. According to this model, an elementary rounded body grows into a filamentous form and then new elementary rounded bodies are developed within this filament and released, but this model has not been adequately substantiated. The cell reproduction of *Mycoplasma capricolum* was studied. The velocity of DNA replication fork progression was about 6 kb/min, which is 10 times slower than that of *Escherichia coli.* The time required for one round of DNA replication accorded with the doubling time. The origin/terminus ratio was 2.0. *M. capricolum* cell morphology was classified into two types, rod and branched. In the ordinary-growth phase, the rod cells accounted for about 90% of the total population, with branched cells comprising the remaining 10%. The proportion of branched cells increased to 90% following inhibition of DNA replication by nucleoside starvation. Transfer of a temperature-sensitive mutant deficient in DNA replication to the restrictive temperature induced an increase in the proportion of branched cells. The rod cells had a regular structure, a fixed cell length, and constrictions in the center. The DNA contents of individual rod cells were distributed with a standard deviation of 0.40 of average. The branched cells had irregular structures and a wide distribution of DNA contents. Counting of viable cells revealed that the cells ceased division upon cell type conversion; however, branched cells maintained a reproductive capacity. A model for the reproduction process is proposed. DNA replication of *Mycoplasma capricolum* starts at a fixed site neighboring the dnaA gene and proceeds to both directions after a short arrest in one direction. The initiation frequency fits to the slow speed of replication fork and DNA content is set constant. The replicated chromosomes migrate to one and three quarters of cell length before cell division to ensure delivery of the replicated DNA to daughter cells. The cell reproduction is based on binary fission but a branch is formed when DNA replication is inhibited. *Mycoplasma pneumoniae* has a terminal structure, designated as an attachment organelle, responsible for both host cell adhesion and gliding motility. Behavior of the organelle in a cell implies coupling of organelle formation to the cell reproduction cycle. Several proteins coded in three operons are delivered sequentially to a position neighboring the previous organelle and a nascent one is formed. One of the duplicated attachment organelles migrates to the opposite pole of the cell before cell division.

More recent studies utilizing highly specific antisera against glycolytic enzymes show that another main division, the "small chromosome" Mycoplasma (the most numerous group of mycoplasmas) is also related closely to the lactic acid bacteria. These findings support that mycoplasmas evolved from walled bacteria by substantial losses of chromosomal material. It is now recognized that closely related bacteria have maintained similar genetic maps, and several lines of evidence suggest that the arrangement of genes in the chromosome may have selective value. It has been observed by different mycoplasmalists that certain mycoplasma species display among their strains a marked heterogeneity in the electrophoretic pattern of their cell proteins or in chromosomal restriction patterns, while other species are quite homogenous. The conserved gene markers are being used to elucidate the basis for this intraspecies variation. Another area of research involves the examination of several important human diseases that are suspected of being caused by bacterial infections but where etiologic agents have not been identified. Recently polymerase chain reaction (PCR) primers have designed for the rapid laboratory diagnosis of tuberculosis. These primers produce specific DNA fingerprints directly from *Mycobacterium tuberculosis* as well as certain other mycobacteria including *M. avium*, *M. intracellulare* complex strains.

It has assayed the time required for completion of replication of a pulse-labeled endonuclease fragment to determine the time for one round of chromosome replication. [32P]damp is used, which can be incorporated into the chromosome. It is tested whether [32P]dAMP taken up into a mycoplasma cell can be rapidly diluted by cold dAMP by monitoring the polymerization process of Okazaki fragments. Mycoplasma cultures were labeled with 16 nM [32P]dAMP for 2 min; then 1 mM cold dAMP was added, and DNA was isolated after various incubation periods and analyzed by denaturing gel electrophoresis. Okazaki fragments were widely distributed in size. The labeling of nascent Okazaki fragments was stopped at 10 s after the addition of cold dAMP, and the sizes of labeled small fragments started to shift. This result showed that [32P]dAMP was available for the pulse-labeling of the chromosomal DNA.

Mycoplasma cultures were pulse-labeled with [32P]dAMP for 2 min, and DNA was isolated after various incubation periods, digested with BamHI, and subjected to FIGE. The radioactivity of each fragment was detected by autoradiography and quantified. The band intensities increased with the incubation period and became saturated at times, which depended on the fragment size. The largest two BamHI fragments, Bm1 and Bm2 were used, because of their separation from other fragments in the gel. The band intensities of Bm1 and Bm2 increased linearly with the incubation period and became saturated at 48.5 and 42.2 min, respectively. The fork velocities were calculated to be 6.4 and 6.0 kb/min, respectively, from the results of Bm1 and Bm2, and the times required for one round of chromosome replication were estimated to be 91 and 97 min, respectively. These values did not depend on the growth phase if the medium was supplemented with the nucleoside mix. The fork velocity in the nonsupplemented cultures was similar to that in the supplemented cultures until the OD600 reached 0.1. However, a marked reduction was observed after the OD600 reached 0.1.

Titrating the origin/terminus copy number ratio monitored the ratio of replicating intermediates in the chromosome molecules. The origin/terminus ratio was estimated by the hybridization method. The origin region has been defined and a DNA fragment expanding in this region was used for titrating the origin. The region, which is replicated last, has not been identified. Therefore, a DNA region expanding on the gyrase gene has used, which was reported to be in the position opposite the origin on the chromosome map. The origin/terminus ratio was estimated to be 2.0. This value did not change through the growth phase, and it did not depend on nucleoside supplementation.

Transmission electron microscopy revealed that the morphology of *M. capricolum* cells was classified into two types, i.e., rod and branched types. Most cells in the ordinary-growth phase had a comparatively regular rod-like structure. A small fraction of cells had an irregular branched structure, with tubes radiating out from the center of the cell body. These cell types could also be distinguished by light microscopy.

The occurrence of branched cells was examined by phase-contrast microscopy. In cultures supplemented with nucleosides, the proportion of branched cells did not change significantly until the OD600 reached 0.17, after which it started to increase, and the branched cells finally accounted for 37% of the total cell number. In cultures without nucleoside supplementation, the proportion increased in the earlier stage. The proportion started to increase after the OD600 reached 0.10, and the branched cells finally accounted for 90% of the total cell number. The starting point of the increase in the proportion of branched cells corresponded to the cessation of DNA synthesis due to nucleoside starvation, with subsequent protein synthesis. These results suggest that nucleoside starvation induced cell type conversion.

Cell type conversion in a temperature-sensitive mutant deficient in DNA replication. It is examined that the cell type conversion of a temperature-sensitive mutant strain in which the elongation reaction of DNA replication is specifically inhibited at the restrictive temperature. Mutant and wild-type cells were grown at 33°C until the OD600 reached 0.05, and then the incubation temperature was shifted to the nonpermissive temperature of 41°C . The OD600 of mutant cells continued to increase for more than 180 min after the temperature shift. The proportion of branched cells started to increase just after the temperature shift, so that they accounted for 54% of the total cell population at 120 min after the temperature shift, while it did not increase significantly in wild-type cultures.

The rod type cells in cultures supplemented with the nucleosides at an OD600 of 0.05 were analyzed by electron microscopy. The average length of rod cells was 0.941 µm, with a standard deviation (SD) of 0.231 µm, and constriction sites were found around the middle in 21.8% (113 of 519) of the rod cells. The average ratio of the distance from the constriction site to the furthest cell pole, relative to the cell length, was 0.576, with an SD of 0.051. The cells stained with DAPI were analyzed by fluorescence microscopy, and the DNA contents in individual cells were estimated from fluorescence intensity. The average DNA content at an OD600 of 0.05 was normalized to 1 U. The SD of DNA content in rod cells was 0.397 U. These results were not considerably different in the other growth stages or in cultures without nucleoside supplementation.

The DNA content of *M. capricolum* in cultures increased with the protein content without nucleoside starvation, and the DNA contents in individual rod cells did not change throughout the growth phase. These observations suggest that the initiation frequency of DNA replication agrees with the rate of increase of protein content, although it is unclear if the frequency can be modified according to the growth conditions, as has been shown for *E. coli*.

The polymerization process of Okazaki fragments in mycoplasma cells has examined in another study. In this study, it was found that a signal migrated rapidly in the denaturing gel, the intensity of which increased with the chase time. It is unlikely that this signal was caused by host exonuclease activity present at the time of lysis, because the signal intensity did not depend on the time from cell lysis to the first centrifugation. The signal was presumably derived from pseudo-Okazaki fragments caused by excision of uracil incorporated into DNA. A homolog of uracil N-glycosylase has also been found in genome analyses of mycoplasmas.

Monitoring of the progression of DNA replication in mycoplasmas has been reported. However, the replication fork velocity has not been studied, because of the difficulty of achieving a synchronous reproductive cycle in mycoplasmas. It is used pulse-labeling coupled with FIGE and examined the fork velocity without synchronization. The absolute velocity of the replication fork in *M. capricolum* was about 10 times slower than that in *E. coli*. This slow progression of DNA replication may be related to the slow growth of mycoplasmas. An absence of replication machinery has not been observed in the genetic components of mycoplasmas. The time for one round of chromosome replication was estimated to be around 94 min. This value roughly corresponded to the doubling time, suggesting that DNA replication occurs in an interval between two cell divisions. This assumption was supported by the origin/terminus ratio. *M. capricolum* cells were morphologically classified into two types, i.e., rod and branched. The rod cells are assumed to be the reproductive form under normal-growth conditions because (i) the majority of cells in normal-growth cultures were rod type, (ii) constrictions were found in 22% of the rod cells, (iii) the rod cells had cell length and DNA content distributions suitable for reproduction by division, (iv) the DNA contents of rod cells did not change during the normal-growth phase, which agrees with the constant increase in the DNA contents of the cultures, and (v) the branched cells cannot be a stage of the ordinary cell division cycle, because the DNA contents of branched cells were not significantly larger than those of rod cells. If the rod cells became branched before division, the DNA contents of branched cells should be significantly larger than that of the rod cells.

Cell type conversion was induced by starvation of nucleosides and by transfer of a temperature-sensitive DNA replication mutant to the nonpermissive temperature. These results suggest that the inhibition of DNA replication with subsequent protein synthesis induced the cell type conversion. In the conversion, cells did not divide; i.e., the increase in CFUs was extensively reduced at the beginning of the conversion, and no anucleate minicells were observed in the microscopic field. It is likely that mycoplasma cells whose division system is ready convert to the branched type when the nonreplicated chromosomal DNA inhibits division.

In ordinary growing cultures, a small proportion of cells were found as the branched type, and the proportion did not depend on the growth stage. Presumably, a small fraction of rod cells occasionally converts to the branched type and then returns to the rod type. The reproductive capability of branched-type cells was confirmed by microplate observation. In ordinary growth, the rod cells divide into two nascent cells. DNA replication occurs in a cell division interval. On the other hand, a cell whose DNA replication is inhibited by nucleoside starvation cannot undergo cell division, and it makes new projections due to the excess potential of cell division. In ordinary growth, a small fraction of rod cells, which have some delay in completion of chromosome replication, also convert to the branched type.

M. mycoides is closely related to *M. capricolum* and appearance of this species is also similar to that of *M. capricolum*. Buxton and Fraser reported that *M. mycoides* develops from an elementary rounded body into a branched and filamentous form, and then new rounded elementary bodies develop within these filaments and are released when they are mature.

Models of cell division into two equivalent cells were also proposed for *Mycoplasma gallisepticum*, *Mycoplasma mobile* and

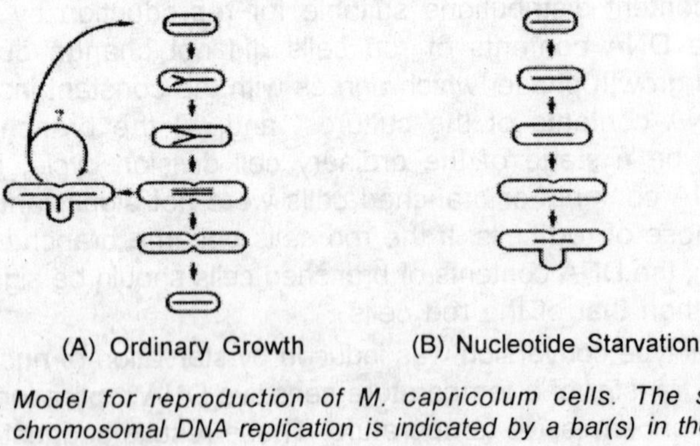

Fig. 3.10. Model for reproduction of M. capricolum cells. The state of chromosomal DNA replication is indicated by a bar(s) in the center of a cell.

Spiroplasma citri, which are also related to *M. capricolum*. The growth phase dependency of the morphological change of *Mycoplasma pneumoniae* is similar to that in *M. capricolum*. Cells of *M. pneumoniae* are spherical in young cultures, become branched and filamentous with the growth phase, and then change to asymmetrical rounded forms in declining cultures. The morphological change in *M. pneumoniae* was suggested to respond to the change in growth medium components. A process similar to that in *M. capricolum* may cause the cell type conversion of *M. pneumoniae*.

A morphological change induced by nucleoside starvation, whereby cells became filamentous with thymine starvation, was also reported for *E. coli*. However, no branching was seen in this case. Akerlund et al. showed that 5% of *E. coli* cells formed branches in nutrient-poor medium when chromosome replication or nucleoid segregation was genetically disturbed. These phenomena may be related to branch formation in mycoplasmas.

Since mycoplasmas lack the peptidoglycan layer, changes in the cytoplasm and cell membrane can be directly reflected in the appearance of the cells. Therefore, branch formation by mycoplasmas is probably coupled with abnormal assembly of proteins that play roles in cell division or maintenance of cell shape. This is supported by the observation that cell type conversion required subsequent protein synthesis. FtsZ protein is known to form a Z ring at the position of septation prior to cell division and to play key roles in cell division in *E. coli* and other walled bacteria. Homologs of the ftsZ gene have been identified in some mycoplasmas, including *M. capricolum*. FtsZ may be related to the branch formation of mycoplasmas. It has been reported that overexpression of FtsZ protein induces a branch at the stalk in Caulobacter crescentus.

Pathogenesis

No other group of procaryotes has been so embroiled in controversy and in establishing a clear pathogenic niche as the mycoplasmas. Their virulence determinants are undeniably complex, and their unique biological properties likely challenge the host differently from typical bacterial pathogens. Also, numerous Mycoplasma species appear to comprise the commensal microbial flora of healthy persons and the association of these mycoplasmas with disease complicates the diagnosis and necessitates extensive and highly specific serologic, nucleic acid, and epidemiologic data. Nonetheless, mycoplasmas by themselves can cause acute and chronic diseases at multiple sites with wide-ranging complications and have been implicated as cofactors in disease.

The earliest reports of mycoplasmas as infectious agents in humans appeared in the 1930s and 1940s. At that time, primary atypical pneumonia was associated with an infectious agent that because of its minute size and innate biological properties unknown at that time, passed through bacteria-retaining filters, resisted penicillin and sulfonamide therapies, and adapted to growth in embryonated eggs and tissue culture cells. Correlations between the etiologic agent of "walking pneumonia" with viruses, L-forms, and pleuropneumonia-like agents were frequent and often misleading. Finally, definitive studies in the early 1960s established *Mycoplasma pneumoniae* as the singular cause of cold agglutinin-associated primary atypical pneumonia. Today *M. pneumoniae* remains an important cause of pneumonia and other airway disorders, such as tracheobronchitis and pharyngitis and is associated with extrapulmonary manifestations, such as hematopoietic, exanthematic, joint, central nervous system, liver, pancreas, and cardiovascular syndromes.

Mycoplasmas are surface parasites of the human respiratory and urogenital tracts. *Mycoplasma pneumoniae* attaches to sialoglycoproteins or sialoglycolipid receptors on the tracheal epithelium via protein adhesins on the attachment organelle. The major adhesin is a 170-kilodalton (kDa) protein, named P1. Hydrogen peroxide and superoxide radicals (O^{2-}) excreted by the attached organisms cause oxidative tissue damage. Pneumonia is induced largely by local immunologic and phagocytic responses to the parasites. Sequelae of *M. pneumoniae* infection (mainly hematologic and neurologic) apparently have an autoimmune etiology. Several fastidious mycoplasmas may act as cofactors in activation of the aquired immunodeficiency syndrome (AIDS). Macrophage activation, cytokine induction, and superantigen properties of some mycoplasmal cell components can be considered as pathogenicity factors.

The confusion associated with *M. pneumoniae*-mediated infections has recurred many times with other mycoplasmas, whose detection in clinical specimens through culture, antibody, or DNA-based testing is frequently dismissed as "only mycoplasmas" even when they appear to be the primary pathogens. Two mycoplasmas commonly found in the urogenital tracts of healthy persons are *Mycoplasma hominis* and *Ureaplasma urealyticum*. However, over the years, the pathogenic roles of these mycoplasmas have been proven in adult urogenital tract diseases, neonatal respiratory infections, and a range of other diseases usually in immunocompromised patients.

Several recent examples illustrate the increasing impact of Mycoplasma species on emerging diseases. *Mycoplasma fermentans* strains were first isolated from the lower genital tract of both adult men and women in the early 1950s, but their role in classic lower genital tract disease has not been established. Reports in the 1970s of *M. fermentans* in the joints of rheumatoid arthritis patients and in the bone marrow of children with leukemia raised expectations for its pathogenic potential; these findings have not been adequately confirmed. Sufficient evidence, however, has accumulated recently to establish an important and emerging role for *M. fermentans* in human respiratory and joint diseases. For example, *M. fermentans* has been detected by specific gene amplification techniques such as polymerase chain reaction (PCR) in the synovial fluid of patients with inflammatory arthritis, but not in the joints of patients with juvenile or reactive arthritis. In two other studies using PCR, *M. fermentans* was identified in the upper respiratory tract of 20% to 44% of both healthy and HIV-infected patients and was associated with acute respiratory distress syndrome in nonimmunocompromised persons.

All mycoplasmas cultivated and identified thus far are parasites of humans, animals, plants, or arthropods. The primary habitats of human and animal mycoplasmas are the mucous surfaces of the respiratory and urogenital tracts and the joints in some animals. Although some mycoplasmas belong to the normal flora, many species are pathogens, causing various diseases that tend to run a chronic course.

Mycoplasma genitalium was detected in the urogenital tract of two patients with nongonococcal urethritis in 1981, but for more than a decade, very little was known about its host distribution and pathogenicity. Early experimental studies established that the organism caused lower genital tract infections in both male and female chimpanzees, with extensive urethral colonization in males and apparent tissue invasion, eventually leading to overt bacteremia. However, the fastidious growth requirements of *M. genitalium* from human hosts severely limited further study until the advent of molecular detection techniques. Specific sequences in the 140 kDa adhesin protein gene of *M. genitalium* were selected as targets in a PCR-based detection assay.

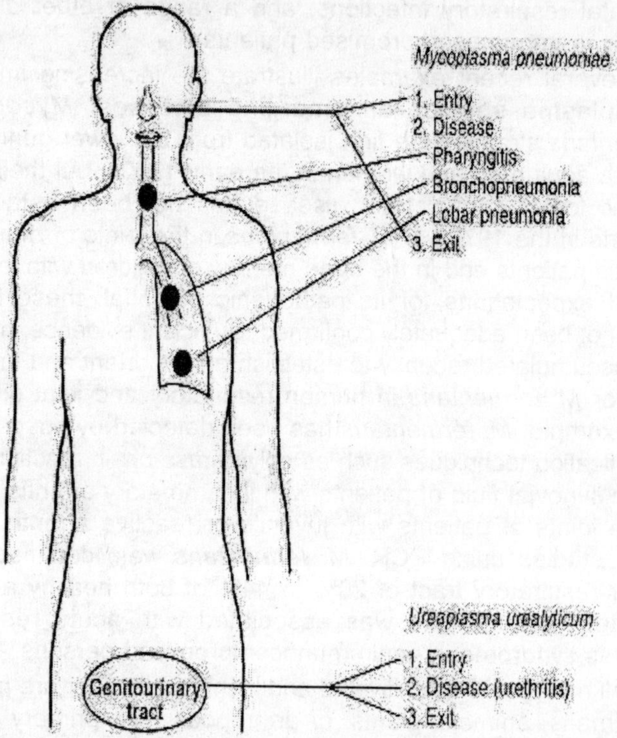

Fig. 3.11. Pathogenesis and disease sites of infection by M. pneumoniae and U. urealyticum.

Subsequent application of these techniques in cases of acute nongonococcal urethritis, not including those of patients colonized or infected with *Chlamydia trachomatis*, has provided mounting evidence for the involvement of *M. genitalium* as an etiologic agent of this disease. Also, *M. genitalium* has been suspected in chronic nongonococcal urethritis and pelvic inflammatory disease.

The discovery in 1988 of *M. genitalium* strains in human nasopharyngeal throat specimens, where they were frequently mixed with strains of *M. pneumoniae*, not only changed dramatically the concept of host distribution of *M. genitalium* but also prompted critical questions about the role of this mycoplasma in human respiratory disease. However, the immunologic cross-reactivity between *M. genitalium* and *M. pneumoniae* and the inability of most conventional diagnostic serologic tests to conclusively identify *M. genitalium* have complicated its delineation in acute human respiratory disease. PCR assays specific for the organism have detected *M. genitalium* in throat specimens of patients infected with HIV-1. However, these probes have not been applied to control groups and patients in outbreaks of acute respiratory disease and/or pneumonia to determine whether *M. genitalium* alone is an etiologic agent in respiratory infections.

M. genitalium has been implicated as an etiologic agent in certain human joint diseases. This clinical correlation began with the observation of a mixed infection of *M. pneumoniae* and *M. genitalium* in synovial fluid specimens of a nonimmunocompromised patient after an acute respiratory infection. A predominant role was not established for either Mycoplasma species in the initial respiratory disease or in the joint manifestations, although evidence to implicate postinfectious autoimmunity to both organisms was described. These findings prompted a PCR assay on synovial fluids from patients with various arthritic syndromes, which presented case reports on two of 13 patients with *M. genitalium* detected in joint fluids.

Another area of emerging mycoplasmal infections concerns immunodeficiency. Although patients with congenital or acquired disorders of antibody production are susceptible to a wide variety of microbial infections, the unique susceptibility of such patients to mycoplasmal infections is a growing concern, especially considering the number of occurrences, the types of mycoplasmas involved, and the difficulties posed in the therapeutic management of such infections. In addition, the increased use of prolonged or permanent immunosuppressive chemotherapy required for patients undergoing tissue or organ transplantation or treatment of various malignant diseases has also increased the risk for mycoplasmal infections from mycoplasmas that are part of the normal human mollicute flora to those acquired through animal contact.

The association between immunodeficiency and mycoplasmal infections was first reported in the mid 1970s in patients with primary hypogammaglobulinemia and infection with *U. urealyticum*,

M. pneumoniae, *Mycoplasma salivarium*, and *M. hominis* that localized in joint tissue, frequently with destructive arthritis. Similar joint infections in hypogammaglobulinemic patients with these mycoplasmal species continue to be reported. Since most of these mollicutes, with the possible exception of *M. pneumoniae*, occur as part of the normal human flora, the origin of such joint infections is considered endogenous. Patients with hypogammaglobulinemia and other antibody deficiencies are also especially susceptible to mycoplasmal infections of the upper respiratory and urinary tracts caused most frequently by *M. pneumoniae* or *U. urealyticum*, respectively.

Mycoplasmal infections following organ transplantation and immunosuppressive chemotherapy were observed in the early 1980s, with both *M. hominis* and *U. urealyticum* reported most often. Although these infections most likely originated from the patient's normal microbial flora, a recent report of donor transmission of *M. hominis* to two lung allograft recipients suggests that donor tissue may be a more important factor in transplant infections than currently recognized.

While patients with antibody defects or those receiving immunosuppressive drugs appear to be the most susceptible to infections with mycoplasmas present in healthy tissues, emerging evidence indicates that contact with other mycoplasmas in the environment is an important hazard. For example, the direct isolation of a feline mycoplasma (*M. felis*) from the joint of a hypogammaglobulinemic patient with septic arthritis was recently reported with suspected transmission occurring through a cat bite 6 months before the onset of arthritis. Other examples include fatal septicemia caused by *M. arginini*, a common animal mycoplasma, from blood and multiple tissue sites in a slaughter house employee who had advanced non-Hodgkin's lymphoma and hypogammaglobulinemia and a septicemic infection with a canine mycoplasma (*M. edwardii*) in a patient with advanced AIDS.

One of the most critical aspects of mycoplasmal infections in immunodeficient patients is the frequent inability to control such infections with appropriate broad spectrum antibiotics. Although the tetracyclines and erythromycins are effective chemotherapeutic agents for many mycoplasmal infections, *M. fermentans* and *M. hominis* strains are usually resistant to erythromycin, and tetracycline-resistant strains of *M. hominis* and *U. urealyticum* have been reported from the lower urogenital tract of patients. However, these antibiotics and most other broad spectrum agents have limited mycoplasmacidal activity in vivo, and their efficacy eventually depends on an intact host immune system to eliminate the mycoplasmas. Most hypogammaglobulinemic patients lack the ability to mount a strong antibody response. Guidelines for managing such mycoplasmal infections in patients with immune defects should include immediate in vitro testing of the isolated mollicute against a wide range of antibiotics; expeditious administration of the antibiotic by the most appropriate route prolonged therapy terminated only if there is no rapid clinical or microbiological response; and possibly administration of intravenous immunoglobulin). Clinical management of mycoplasmal infections in transplant patients is more difficult since immunoglobulins may enhance graft or organ rejection. In the absence of suitable mycoplasmacidal chemotherapeutic agents, vigorous and sustained chemotherapy with the most active antibiotic is the current method of choice.

Mechanisms of Pathogenicity

Many mycoplasmal pathogens exhibit filamentous or flask-shaped appearances and display prominent and specialized polar tip organelles that mediate attachment to host target cells. These tip structures are complex, composed of a network of interactive proteins, designated adhesins, and adherence-accessory proteins. These proteins cooperate structurally and functionally to mobilize and concentrate adhesins at the tip and permit mycoplasmal colonization of mucous membranes and eucaryotic cell surfaces, probably through host sialoglycoconjugates and sulfated glycolipids. It appears that mycoplasmal cytadherence-related proteins represent a superfamily of genes and proteins that have been conserved through horizontal gene transfer from an ancestral gene family. This protein network resembles a specialized cytoskeleton-like apparatus, which may represent the precursor to mammalian cytoskeletal and extracellular matrix-like complexes. Other Mycoplasma species lack distinct tip structures yet are capable of cytadherence, and they may use related genes or proteins or alternative mechanisms of surface parasitism.

The family of mycoplasmal genes and proteins involved in cytadherence has been studied most extensively in *M. pneumoniae*. Noncytadhering phenotypes that arise through spontaneous mutation at high frequency have been categorized into mutant classes on the basis of distinct protein profiles. These noncytadhering mycoplasmas cannot synthesize specific cytadherence-related proteins or are unable to stabilize them at the tip organelle, which leads to abnormal anatomical tip structures and avirulence. The reappearance of the implicated proteins, restoration of structurally and functionally intact tips, and return of full infectivity accompany spontaneous reversion to the cytadhering phenotype. Similar cytadherence-related genes and proteins have been reported for *M. genitalium* on the basis of biochemical, immunologic, and genetic analyses. Furthermore, striking similarities exist in the order of operons that comprise the cytadherence-related genes and the organization of these genes within each operon of *M. pneumoniae* and *M. genitalium*. These similarities reinforce the unexpected coisolations of *M. genitalium*, along with *M. pneumoniae*, from the nasopharyngeal throat swabs of patients with acute respiratory diseases and from synovial fluids of patients with arthritis. The isolation of *M. pneumoniae* from the human urogenital tract further suggests that these mycoplasmas have evolved parasitic strategies that include overlapping tissue tropisms as determined by the genetic and chemical relatedness of their cytadherence genes and proteins. An interesting feature of specific *M. pneumoniae* and *M. genitalium* adhesins is their multiple gene copy nature. Although only one full-length copy of the adhesin structural genes exists in adhesin-related operons, precise regions of these adhesin genes are detected as single genomic copies, while other regions occur as closely homologous, but not identical, multiple copies. In other words, multiple truncated and sequence-related copies of the adhesin genes are dispersed throughout the genome, which could generate adhesin variation through homologous recombination. Consistent with this possibility is the existence of restriction fragment length polymorphisms in the adhesin genes of human clinical isolates of *M. pneumoniae* and *M. genitalium*, reflected by sequence divergence in the multiple-copy regions of the adhesin genes. It appears that a repertoire of partial adhesin-related gene regions serves as a reservoir to regulate the structural and functional properties of mycoplasmal adhesins through recombination events, which may lead to circumvention of the host immune response. Mechanisms of phase and antigenic variation are likely to occur in which mycoplasmal adhesins exhibit altered specificities and affinities, as determined by the organization of constant and variable adhesin gene sequences. Therefore, despite their small genomes, pathogenic mycoplasmas facilitate DNA rearrangements through repetitive gene sequences, thus promoting genetic diversity and maximizing the coding potential of their limited genomes. The immunodominant epitopes of the mycoplasmal adhesins appear not to be identical to the adherence-mediating domains. The latter are in part encoded by single copy regions of the adhesin genes and are highly conserved, which reinforces their essential role in mycoplasmal recognition of host cell receptors and colonization. Host immunoresponsiveness directed at the noncytadherence-mediating variable regions is unlikely to generate effective cytadherence-blocking antibodies, which may in part clarify the observed high reinfection rates of patients. Thus, the grouping of clinical isolates of *M. pneumoniae* into two categories, on the basis of sequence divergence in the multiple-copy regions of the adhesin gene, along with the immune status of the population, may explain the epidemiologic patterns of *M. pneumoniae*.

Another characteristic of the cytadherence-related proteins is their proline-rich composition, which markedly influences protein folding and binding. Several reports have established the importance of these proline-rich domains in mycoplasmal cytadherence and virulence and recent evidence further suggests that mycoplasmal peptidylprolyl isomerases, i.e., cyclophilins, are critical in regulating the conformation and function of the mycoplasmal cytadherence-related tip organelle, colony morphology, and growth. In addition to this proline-rich property, one of the most unusual features of the adhesins is their extensive sequence homology to mammalian structural proteins. This molecular mimicry is especially interesting since it has been suggested for decades that mycoplasmas provoke an antiself response that triggers immune disorders, although the basis for the induction has been elusive. Patients with documented *M. pneumoniae* respiratory infections demonstrate seroconversion to myosin, keratin, and fibrinogen and exhibit extrapulmonary manifestations, such as exanthems and cardiac abnormalities. Furthermore, a classic example of bacteria-mediated autoimmune disorders is the development of acute rheumatic fever following streptococcal infection. Antistreptococcal antibodies reactive against ahelical coiled-coil regions of the M protein crossreact with heart myosin, tropomyosin, and mycoplasmal adhesins. In the latter case, these mycoplasmal adhesins exhibit amino acid sequence homologies with human CD4 and class II major histocompatibility complex lymphocyte proteins, which could generate autoreactive antibodies and trigger cell killing and immunosuppression. Also, mycoplasmas may serve as B-cell and T-cell mitogens and induce autoimmune disease through the activation of antiself T cells or polyclonal B cells. The multiorgan protean manifestations of mycoplasmal infections in humans are consistent with the pathogenesis of autoimmunity. Furthermore, the ability of mycoplasmas to induce a broad range of immunoregulatory events, mediated by cytokine production and direct effects on macrophages, B and T cells, and glial cells, is evidence that mycoplasmas possess the attributes of primary mediators of pathogenesis. For example, cytokine production and lymphocyte activation may either minimize disease through the activation of host defense mechanisms or exacerbate disease through lesion development. Also, a superantigen derived from *Mycoplasma arthritidis*, a mycoplasma pathogenic for rodents, induces arthritis and chronic disease manifestations. It has been suggested that

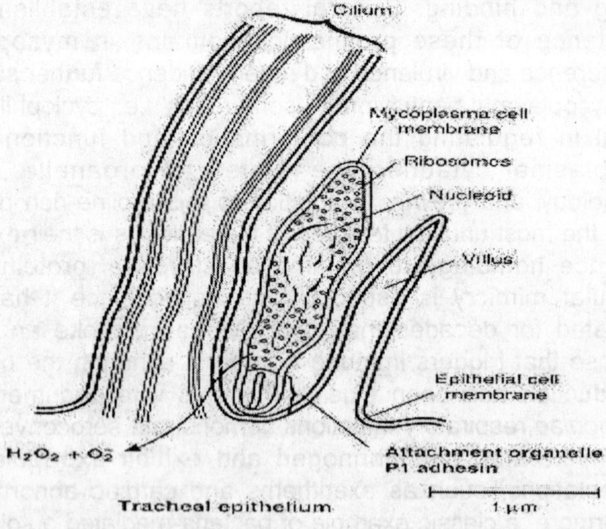

Fig. 3.12. Schematic presentation of a M. pneumoniae organism attaching to the surface of the ciliary tracheal epithelium, as seen by electron microscopy of a thin section.

related superantigen-like molecules may exist in mycoplasmas of human origin triggering autoimmune and other inflammatory pathologies.

It appears that cytadherence is the initial step in the virulence process of pathogenic mycoplasmas and precedes a spectrum of subtle or overt host cell responses. In specific instances, distinct cytopathology correlates with the infecting Mycoplasma species, the number of adherent mycoplasmas, the length of coincubation, the induction of proinflammatory cytokines, and the age and immune status of the patient. For example, the exacerbation of clinical syndromes may correlate with a history of mycoplasmal infection as observed in patients with recurrent *M. pneumoniae* exposures. Also, the elevated expression of proinflammatory cytokines associated with mycoplasmal disease pathogenesis may coincide with the intensity of the symptoms. In other cases, chronic disease or no obvious signs or symptoms of disease accompany mycoplasmal infection.

Other biological properties of mycoplasmas have been implicated as virulence determinants and include:

1. Generation of hydrogen peroxide and superoxide radicals by adhering mycoplasmas, which induces oxidative stress, including host cell membrane damage.
2. Competition for and depletion of nutrients or biosynthetic precursors by mycoplasmas, which disrupts host cell maintenance and function.
3. Existence of capsule-like material and electron-dense surface layers or structures, which provides increased integrity to the mycoplasma surface and confers immunoregulatory activities.
4. High-frequency phase and antigenic variation, which results in surface diversity and possible avoidance of protective host immune defenses.
5. Secretion or introduction of mycoplasmal enzymes, such as phospholipases, ATPases, hemolysins, proteases, and nucleases into the host cell milieu, which leads to localized tissue disruption and disorganization and chromosomal aberrations.
6. Intracellular residence, which sequesters mycoplasmas, establishes latent or chronic states, and circumvents mycoplasmicidal immune mechanisms and selective drug therapies.

Whether pathogenic mycoplasmas enter and survive within mammalian cells has been debated for many years. Consistent with this possibility, mycoplasmas exhibit limited biosynthetic capabilities; are highly fastidious and dependent upon the host microenvironment and complex culture medium for growth; have been observed in intimate contact with mammalian cell surfaces and within target cells; may be capable of initiating fusion with host cells through their cholesterol-containing unit membranes; and survive long-term recommended antimicrobial treatment in humans and tissue cultures. Recent sightings of intact mycoplasmas throughout the cytoplasm and the perinuclear regions of tissue cells from infected patients and in cell cultures, along with evidence that mycoplasmas are capable of long-term intracellular survival and replication in vitro, offer an additional dimension to the pathogenic potential of mycoplasmas.

Since no obvious single or group of mycoplasmal properties inextricably correlates with disease manifestations, the proof that mycoplasmas are card-carrying pathogens necessitates thorough and highly specific microbiological, epidemiologic, and diagnostic criteria; detailed descriptions of biochemical, genetic, and immunologic characteristics that distinguish virulent and avirulent mycoplasmas; and reproducibility of the symptoms of disease in experimental animal models or in the natural spread of infection among susceptible populations. The portfolio of available evidence concerning mycoplasma-mediated disease pathogenesis is limited. These scientific shortcomings precipitate misconceptions concerning mycoplasmas as singular agents of infectious diseases, as putative cofactors in the progression of other diseases, and as universal contaminants of cell cultures. Clearly, multiple pathways of interaction with target cells appear to be the modus operandi of the Mycoplasma species.

Because attachment of *M. pneumoniae* and *M. genitalium* is affected by pretreatment of the host cells with neuraminidase, sialoglycoproteins and/or sialoglycolipids of the host cell membrane appear to be receptor sites for these mycoplasmas. There is evidence that several *M. pneumoniae* membrane proteins act as adhesins and that they have high affinity for the specific receptors for *M. pneumoniae* on host cells. Monoclonal antibodies to one of these proteins, protein P1 (molecular weight, 170,000 daltons), inhibit attachment of the parasite. Ferritin labeling of the antibodies has shown that P1 concentrates on the tip structure of the mycoplasma, a finding that further supports the notion that the tip serves as an attachment organelle.

The results obtained with *M. pneumoniae* were essentially duplicated recently with *M. genitalium* and showed that in this organism, which closely resembles *M. pneumoniae* morphologically and physiologically, a major adhesin protein, named MgPa, is clustered at the tip organelle. The genes of the major adhesins of *M. pneumoniae* (P1) and of *M. genitalium* (MgPa) were cloned and sequenced, allowing the characterization of these proteins. The two adhesins are alike in many respects and in fact contain extensive areas of homology, as expressed also by shared epitopes. These two proteins may be the product of an ancestral gene that underwent a horizontal gene transfer event.

The nature of the toxic factors that damage the mucosal surfaces infected by mycoplasmas is still unclear. Toxins are rarely found in mycoplasmas. Consequently, researchers considered whether the end products of mycoplasma metabolism were responsible for tissue damage. Hydrogen peroxide (H_20_2), the end product of respiration in mycoplasmas, has been implicated as a major pathogenic factor ever since it was shown to be responsible for the lysis of erythrocytes by mycoplasmas in vitro; however, the production of H_2O_2 alone does not determine pathogenicity,

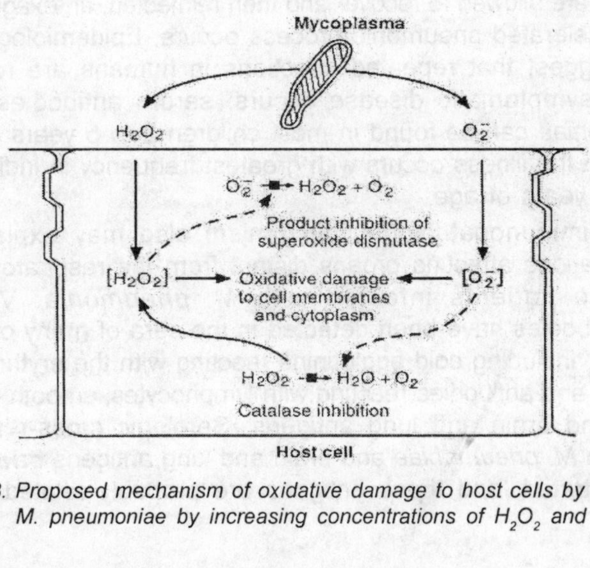

Fig. 3.13. Proposed mechanism of oxidative damage to host cells by adhering M. pneumoniae by increasing concentrations of H_2O_2 and O_2.

as the loss of virulence in *M. pneumoniae* is not accompanied by a decrease in H_2O_2 production. For the H_2O_2 to exert its toxic effect, the mycoplasmas must adhere closely enough to the host cell surface to maintain a toxic, steady-state concentration of H_2O_2 sufficient to cause direct damage, such as lipid peroxidation, to the cell membrane. The accumulation of malonyldialdehyde, an oxidation product of membrane lipids, in cells exposed to *M. pneumoniae* supports this notion. Moreover, *M. pneumoniae* inhibits host cell catalase by excreting superoxide radicals (O_2^-). This would be expected to further increase the accumulation of H_2O_2 at the site of parasite-host cell contact.

There is evidence that both organism-related and host-related factors are involved in the pathogenesis of mycoplasma infections. Mycoplasmas activate macrophages, and induce cytokine production and lymphocyte proliferation; the rat pathogen, *Mycoplasma arthritidis*, produces a potent superantigen. Thus, in the case of *M. pneumoniae*, the host may be largely responsible for the pneumonia by mounting a local immune response to the parasite. Syrian hamsters inoculated intranasally with *M. pneumoniae* show patchy bronchopneumonic lesions consisting of infiltration of mononuclear cells. The ablation of thymic function before the experimental infection prevents development of the characteristic pulmonary infiltration, but lengthens the period during which the organisms may be isolated from the lungs. When thymic animals are allowed to recover and then reinfected, an exaggerated and accelerated pneumonic process occurs. Epidemiologic data also suggest that repeated infections in humans are required before symptomatic disease occurs: serum antibodies to *M. pneumoniae* can be found in most children 2 to 5 years of age, although the illness occurs with greatest frequency in individuals 5 to 15 years of age.

An immunopathologic mechanism also may explain the complications affecting organs distant from the respiratory tract in some patients infected with *M. pneumonia*. Various autoantibodies have been detected in the sera of many of these patients, including cold agglutinins reacting with the erythrocyte I antigen, and antibodies reacting with lymphocytes, smooth muscle cells, and brain and lung antigens. Serologic cross-reactions between *M. pneumoniae* and brain and lung antigens have been demonstrated, and these antigens are probably related to the glycolipids of *M. pneumoniae* membranes, which are also found in most plants and in many bacteria. Clearly, host reaction varies markedly, as only about half of the patients develop cold agglutinins and complications are rare, even among individuals with anti-tissue globulins.

Diseases

There are many different species of Mycoplasma. *Mycoplasma pneumonia* as well as at least 7 other mycoplasma species has been linked as a direct cause or significant co-factor to many chronic diseases. These include *Mycoplasma pneumonia*, incognitos, penetrans, pirum, genitalium, fermentans, salivarium, hominis and *Ureaplasma urealyticum*. Scientists have been studying the relationship between Mycoplasma and different diseases for several decades. Today, over 100 documented species of mycoplasmas have been recorded to cause various diseases in humans, animals, and plants. A list of such conditions includes cancer, auto-immune diseases (arthritis, lupus, scleroderma, multiple sclerosis) ALS, chronic fatigue syndrome and candidiasis. Mycoplasma are found to infect females 4 times more often than males, which just happens to be similar to the incidence rate in rheumatoid arthritis, fibromyalgia, Chronic Fatigue and other related disorders.

Mycoplasmas, unlike viruses, can grow in tissue fluids (blood, joint, heart, chest and spinal fluids) and can grow inside any living tissue cell without killing the cells, as most normal bacteria and viruses will do. Mycoplasmas are frequently found in the oral and genito-urinary tracts of normal healthy people and are found to infect females four times more often than males, which just happen to be the same incidence rate in rheumatoid arthritis, fibromyalgia, Chronic Fatigue and other related disorders. Mycoplasmas are parasitic in nature and can attach to specific cells without killing the cells and thus their infection process and progress can go undetected. In some people the attachment of mycoplasmas to the host cell acts like a living thorn; a persistent foreign substance, causing the host's immune defense mechanism to wage war. This allergic type of inflammation often results in heated, swollen, and painful inflamed tissues, like those found in rheumatoid diseases, fibromyalgia and many other autoimmune disorders like lupus and MS, Crohn's and others. In such cases the immune system begins attacking itself and/or seemingly healthy cells. Some species of mycoplasmas also have the unique ability to completely evade the immune system. Once they attach to a host cell in the body, their unique plasma and protein coating can then mimic the cell wall of the host cell and the immune system cannot differentiate the mycoplasma from the body's own host cell.

Mycoplasmas are parasitic in nature because they rely on the nutrients found in host cells including cholesterol, amino acids, fatty acids and even DNA. They especially thrive in cholesterol rich and arginine-rich environments. Mycoplasmas can generally be found in the mucous membrane in the respiratory tract. They need cholesterol for membrane function and growth, and there is an abundance of cholesterol in the bronchial tubes of the respiratory tract. Once attached to a host cell, they then begin competing for nutrients inside the host cells. As nutrients are depleted, then these host cells can begin to malfunction, or even change normal functioning of the cell, causing a chain reaction with other cells (especially within the immune and endocrine systems). Mycoplasmas can even cause RNA and DNA mutation of the host cells and have been linked to certain cancers for this reason. Mycoplasmas can also invade and live inside host cells, which evade the immune system, especially white blood cells. Once inside a white blood cell, mycoplasmas can travel throughout the body and even cross the blood/brain barrier, and into the central cervous system and spinal fluid.

Members of the genus Mycoplasma are obligate parasites and pathogens of a wide range of mammalian and avian hosts and tend to be adapted to a specific host. Whilst this may result in a commensal type relationship, some species cause serious disease in domestic animals – for example contagious bovine pleuropneumonia (*Mycoplasma mycoides* subsp. *mycoides*). Identification of species can sometimes be difficult due to the presence of more than one species. Mycoplasmas are most unusual self-replicating bacteria, possessing very small genomes, lacking cell wall components, requiring cholesterol for membrane function and growth, using UGA codon for tryptophan, passing through "bacterial-retaining" filters, and displaying genetic economy that requires a strict dependence on the host for nutrients and refuge. In addition, many of the mycoplasmas pathogenic for humans and animals possess extraordinary specialized tip organelles that mediate their intimate interaction with eucaryotic cells. This host-adapted survival is achieved through surface parasitism of target cells, acquisition of essential biosynthetic precursors, and in some cases, subsequent entry and survival intracellularly. Misconceptions concerning the role of mycoplasmas in disease pathogenesis can be directly attributed to their biological subtleties and to fundamental deficits in understanding their virulence capabilities. In this review, we highlight the biology and pathogenesis of these procaryotes and provide new evidence that may lead to increased appreciation of their role as human pathogens.

M. pneumoniae is an extracellular pathogen that adheres to the respiratory epithelium by a specialized terminal protein attachment factor. This adherence protein interacts specifically with neuraminic acid residues on the epithelial cell surface. Ciliastasis occurs following attachment and then destruction of the superficial layer of epithelial cells. Destruction is due to release of hydrogen peroxide and superoxide anion.

Mycoplasma produce primary atypical pneumonia. This disease can range from subclinical to bronchopneumonia, often with a gradual onset and mild to moderate severity. A long convalescence (4-6 weeks) and several possible complications (CNS, cardiac) follow acute disease. *U. urealyticum* is responsible for one form of nongonococcal urethritis. These bacteria are surface parasites that produce locally high concentrations of toxic metabolites (hydrogen peroxide plus inhibition of cell catalase, ammonia). They may also produce ciliostasis and altered host cell metabolism. The host response is one of pneumonia via infiltration of mononuclear cells. Immunopathologic mechanisms and autoantibodies may also result due to cross reactivity with the Mycoplasma membrane glycolipids. Host defenses are not well characterized but probably involve both humoral and cell mediated responses. Mycoplasma affect a specific age distribution (5-9 year olds) and represent 8-15% of all pneumonias in school age children. Disease occurs worldwide, is endemic in some areas and is spread by close personal contact (schools, families). *U. urealyticum* is sexually acquired. Symptomology is equivocal and ambiguous; diagnosis is dependent on isolation.The bacteria can be grown using complex media with penicillin. A fluorescent antibody stain can be used to confirm the identification. For controlling the diseases, it is necessary to avoid the contacts, if possible. There is no single vaccine available for immunological control. Chemotherapeutic, Tetracycline or chloramphenicols are effective.

The human diseases caused by mycoplasmas.

Disease or Symptom	*Agent*	*Host*
Primary atypical pneumonia	Mycoplasma pneumoniae	Man
Non-gonococcal urethritis	Mycoplasma genitalium	Man
Non-gonococcal urethritis	Ureaplasma urealyticum	Man
Stillbirth	Mycoplasma hominis	Man
Spontaneous abortion	Mycoplasma hominis	Man
Infertility	Mycoplasma hominis	Man

Due to the small size and the lack of a cell wall, this microorganism is capable of infecting a great number of cells in any part of the body and live as a parasite on the surface of the cells. Mycoplasmas can become a parasite in plants, insects, animals, and humans, and can trigger different diseases. Once the Mycoplasma becomes a parasite in the cell, morphologic and physiologic changes are developed and it takes on the appearance of various diseases such as pneumonia, urethritis, pyelonephritis, as well as arthritis, lupus and other immune diseases. In cancer, the tumor cells infected by Mycoplasma are more susceptible to spreading which contributes to metastases in different parts of the body. This is also true with Chronic Fatigue Syndrome (especially when the symptoms are severe), Multiple Sclerosis, and ALS (Amyotrophic Lateral Sclerosis).

A list of conditions which may be caused or influenced by Mycoplasma or for which Mycoplasma becomes an opportunistic infection include:

- AIDS/HIV
- ALS
- arthritis
- Asthma
- Bronchitis
- Cancer
- Chronic fatigue syndrome
- Chronic nongonococcal urethritis

Chronic pelvic inflammatory disease
Crohn's
Diabetes
Endocrine disorders
Eye and ear disorders and infections
Fibromyalgia
Gingivitis and periodontal disease
Heart disease
Infertility
Irritable bowel syndrome
leukemia
lupus
Multiple sclerosis
Pneumonia
Psoriasis
Scleroderma
TMJ disorders
Urinary tract infections

Nongonococcal Urethritis and Salpingitis

Growing evidence suggests that *Ureaplasma urealyticum* causes nongonococcal urethritis in men free of *Chlamydia trachomatis*, an established agent of nongonococcal urethritis. The wide occurrence of *U. urealyticum* in sexually active, symptom-free adults hampers research in this field. Evidence is based primarily on the production of nongonococcal urethritis symptoms in ureaplasma-free and chlamydia-free volunteers by intraurethral inoculation of *U. urealyticum* and on a report that this disease could be cured in a chlamydia-free man only when he and his partner were treated simultanously with tetracycline, which eliminated *U. urealyticum* from both. *Ureaplasmas* have also been associated with chorioamnionitis, habitual spontaneous abortion, and low-weight infants. *Mycoplasma hominis*, a common inhabitant of the vagina of healthy women, becomes pathogenic once it invades the internal genital organs, where it may cause pelvic inflammatory diseases such as tubo-ovarian abscess or salpingitis.

It has been suggested that *Mycoplasma genitalium*, isolated in 1981 from the urethral discharge of two homosexual men, may account for the tetracycline-responsive, nongonococcal urethritis cases in which chlamydias and ureaplasmas cannot be isolated (about 20 percent of all cases). However, *M. genitalium* is so fastidious that very few clinical isolates have so far been made on the best mycoplasma medium available. Only the recent application of specific PCR amplification of the organism's DNA in clinical specimens has provided experimental proof for the relative prevalence of *M. genitalium* in the human urogenital tract and its apparent role in male urethritis.

Nonetheless, mycoplasmas by themselves can cause acute and chronic diseases at multiple sites with wide-ranging complications and have been implicated as cofactors in disease. Recently, mycoplasmas have been linked as a cofactor to AIDS pathogenesis and to malignant transformation, chromosomal aberrations, the Gulf War Syndrome, and other unexplained and complex illnesses, including chronic fatigue syndrome, Crohn's disease, and various arthritides. Even with mounting evidence of their pervasive and pathogenic potential, mycoplasmas still evoke the image of a group of obscure or impotent microorganisms. Yet they are evolutionarily advanced prokaryotes and their elite status as "next generation" bacterial pathogens necessitates new paradigms in fully understanding their disease potential.

AIDS

The role of mycoplasmas in accelerating the progression of AIDS could not have begun under more baffling and circuitous conditions. The question of whether mycoplasmas act as co-factors in the development of AIDS has attracted much attention recently. Several mycoplasms have so far been incriminated: *M. fermentans*, considered until recently a relatively rare mycoplasma of the human urogenital tract, and *M. penetrans*, a newly-discovered human mycoplasma isolated from several AIDS patients. *M. pirum*, a mycoplasma of an unknown host, has been recently isolated from the blood of a few AIDS patients. While, in vitro studies show that these mycoplasmas may markedly enhance pathogenicity of the human immunodeficiency virus, the possibility that the mycoplasmas may simply represent opportunistic agents found in high frequency in patients with AIDS, cannot be ruled out. Yet on the whole, with the increasing incidence of immunocompromised patients (due to AIDS, organ transplantation, etc.) evidence is accumulating for invasion of tissues and the intracellular location of some mycoplasmas, notably *M. fermentans* and *M. penetrans*. Extragenital infections by urogenital mycoplasmas are rather common in neonates, immunosuppressed and/or hypogammaglobulinemic patients; clinical symptoms are expressed frequently as arthritis.

A virus-like agent that arose through transfection of NIH 3T3 cells with DNA from Kaposi sarcoma tissues of AIDS patients was later shown to be M. fermentans. The spotted history of *M. fermentans* in rheumatoid arthritis and leukemia and its frequent contamination of cell cultures, along with its contemporary link to AIDS, have been considerable impediments to overcome in its elevation to pathogenic status. However, careful and convincing independent studies by several laboratories have implicated *M. fermentans* as a cause of systemic infections and organ failure in AIDS patients. The isolation of *M. fermentans* from blood and urine samples of HIV-infected persons, its detection by PCR and immunohistochemistry in multiple tissue sites at various stages of AIDS, and its ability to stimulate CD4+ lymphocytes and other immunomodulatory activities implicate this Mycoplasma species as a cofactor in AIDS. Consistent with this possibility, *M. fermentans* has been shown to act synergistically with HIV to enhance cytopathic effects on human CD4+ lymphocytes. Coincident with these studies, a new Mycoplasma species, *Mycoplasma penetrans*, also has emerged as a potential cofactor in AIDS progression. Its isolation almost exclusively from the urine of HIV-infected patients, the extraordinarily high prevalence of antibodies against this mycoplasma in HIV-infected patients and not in HIV-seronegative persons, and its capacity to invade target cells and activate the immune system of HIV-infected patients at various stages of disease correlate with a synergistic role with HIV. Other mycoplasmas, including *M. genitalium* and *Mycoplasma pirum*, have also been isolated from AIDS patients and implicated as potential cofactors. However, the proposed role of mycoplasmas as infectious agents and cofactors in AIDS-related disorders still remains a hypothesis without definitive proof. If cofactors of HIV are essential to the development of late stages of HIV-mediated disease, mycoplasmas possess all the prerequisite properties of the consummate helper. Their ability to establish covert or overt chronic and persistent infections with concomitant activation of the immune system, stimulation of cytokine production, and induction of oxidative stress correlate with increased HIV replication and disease progression.

Malignant Transformation

As early as the mid-1960s, mycoplasma-infected cell lines were associated with chromosomal aberrations, altered morphologies, and cell transformation. These abnormal oncogenic cell traits continued even after the apparent elimination of mycoplasmas, and evidence implied increased tumorigenicity of these transformed cells in animals. This issue has been revisited in studies demonstrating that longterm, persistent mycoplasmal infection of mouse embryo cells initiated a multistage cellular process that resulted in irreversible cell transformation, karyotypic alterations, and tumorigenicity in nude mice.

Gulf War Syndrome

One of the most controversial current medical issues is whether the multiple acute and chronic symptoms found in veterans of the Persian Gulf War were caused by chemical exposure, infectious agents, or psychological problems, or whether a Gulf War Syndrome exists at all. The clinical illness comprises a collection of symptoms, including chronic fatigue, joint pain, headaches, and skin rashes. One study suggests that pathogenic mycoplasmas are responsible for a large number of cases among veterans, on the basis of DNA hybridization and the responsiveness of veterans to prolonged antibiotic treatment. Even though the experimental evidence is sparse and incomplete and well-controlled and detailed studies by independent laboratories are needed, if the Gulf War Syndrome has infectious causes, mycoplasmas with their requisite biological credentials are potential candidates.

Crohn's Disease

Several epidemiologic studies correlate respiratory infections with exacerbation of Crohn's disease and other chronic inflammatory bowel diseases. Seroconversions to specific viral or *M. pneumoniae* antigens accompany acute onset gastrointestinal symptoms in patients with these diseases. As indicated earlier, mycoplasmas can elicit pleiotropic immune responses and are difficult to eliminate in patients despite appropriate antibiotic treatment. Steroid therapy to control gastrointestinal symptoms in these patients, along with the multifaceted biological properties associated with pathogenic mycoplasmas, may precipitate the onset of acute exacerbations of chronic inflammatory bowel disease.

Rheumatoid Arthritis and Other Human Arthritides

The occurrence of various *Mycoplasma* and *Ureaplasma* species in joint tissues of patients with rheumatoid arthritis, sexually transmitted reactive arthritis, and other human arthritides can no longer be ignored. A clinical trial of longterm (6 to 12 months) antibiotic (doxycycline) therapy before cartilage destruction might prove beneficial in managing such frequent and often debilitating infections.

Extensive clinical and microbiological evidence indicates that mycoplasmas alone can elicit a spectrum of illness for which no other agents are incriminated. The eradication of these pathogenic mycoplasmas from various tissue sites requires an intact and functional immune system, although persons with fully competent immune systems may have difficulty eliminating mycoplasmas, even with recommended prolonged drug therapy. Nonetheless, mycoplasmas are still viewed as subordinates to other infectious agents and are relegated to a category of commensals that unwittingly cause disease in patients whose immune systems offer little resistance to microbial stress and overload.

The fundamental importance of mycoplasmas in specific diseases of humans, animals, insects, and plants is irrefutable, and their unique biological properties are consistent with their intimate association with host target cells. These remarkable bacteria must continue to receive the scientific attention of mycoplasmologists, cell culturists, clinicians, immunologists, and DNA sequencers who most recently are compiling extensive databases that may eventually dissect every approachable mycoplasmal element that defines their biological and genetic being. Nonetheless, mycoplasmas remain mysterious and enigmatic, and the available data and proposed hypotheses that correlate mycoplasmas with disease pathogenesis range from definitive, provocative, and titillating to inconclusive, confusing, and heretical. Controversy seems to be a recurrent companion of mycoplasmas, yet good science and openmindedness should overcome the legacy that has burdened them for decades.

Minocycline in Rheumatoid Arthritis

In two recently-published independent randomized trials, rheumatoid arthritis patients were treated with 100 mg of oral minocycline twice daily or a placebo for a period of 26 weeks. In the minocycline group, more minocycline-treated patients than placebo showed greater than 75% improvement in swollen joint count, tender joint count and in clinical parameters such as serum C-reactive protein (CRP) level and erythrocyte sedimentation rate (ESR). In these studies, the intergroup differences were statistically significant for these findings and the mean changes over time revealed continual improvement in the minocycline-treated patients during the entire period of both studies.

This and other presently-available data on minocycline therapy in rheumatoid arthritis suggest that such treatment may be considered along with disease-modifying anti-rheumatic drugs such as methotrexate, sulfasalazine, gold salts and hydroxychloroquine. However, additional clinical research is necessary to document the long-term efficacy of minocycline in the decreased progression of joint destruction. We believe that such long-term study about the efficacy of minocycline should be conducted on patients who are positive for mycoplasma and chlamydia genome (since we detect the chlamydia trachomatis genome in blood and joint fluid of 20% of patients with rheumatoid arthritis) and not by random selection of arthritis patients. Such selection or comparison between mycoplasma- and chlamydia-positive patients with mycoplasma- and chlamydia-negative individuals may further increase the clinical efficacy of minocycline or doxycycline in future double-blind placebo studies.

The eradication of the pathogenic mycoplasmas from blood and various tissue sites requires an intact functional immune system, which most patients with chronic illnesses do not possess. Therefore, immune enhancement strategies along with prolonged drug therapy may help to eliminate mycoplasma from the human body.

Host Defenses

Infection with *M. pneumoniae* induces the development of serum antibodies that fix complement, inhibit growth of the organism and lyse the organism in the presence of complement. Generally, the first antibodies produced are of the IgM class, whereas later in convalescence the predominant antibody is IgG. Secretory IgA antibodies also develop and appear to be important in host resistance. The first infection in infancy usually is asymptomatic and generates a brief serum antibody response. Recurrent infections generate a more prolonged systemic antibody response and increasing numbers of circulating antigen-responsive lymphocytes. By late childhood, clinically apparent lower respiratory disease, including pneumonia, becomes more common. Therefore, mycoplasma respiratory disease manifestations appear to vary, depending on the state of local and systemic immunity at the time of reinfection. One hypothesis is that local immunity mediates resistance to infection and that systemic immunity contributes substantially to the pulmonary and systemic reaction characteristic of *M. pneumoniae* pneumonia.

The relative importance of humoral and cell-mediated immunity in resistance to respiratory mycoplasma infections is still unclear. For many mycoplasma infections, such as bovine pleuropneumonia, resistance can be transferred with convalescent-phase serum, but this may not be true for all mycoplasma respiratory diseases. For example, resistance of rats to pulmonary disease induced by *M. pulmonis* can be transferred only with spleen cells obtained from previously infected animals. Although IgA antibody may be important in resistance to mycoplasmas, other factors seem to be involved in resistance to pulmonary disease, and these factors may not be the same for all mycoplasma infections.

Epidemiology

One of the most puzzling features of *M. pneumoniae* pneumonia is the age distribution of patients. In a survey conducted between 1964 and 1975 of more than 100,000 individuals in the Seattle area, the age-specific attack rate was highest among 5- to 9-year-old children. Rates of *M. pneumoniae* pneumonia in the

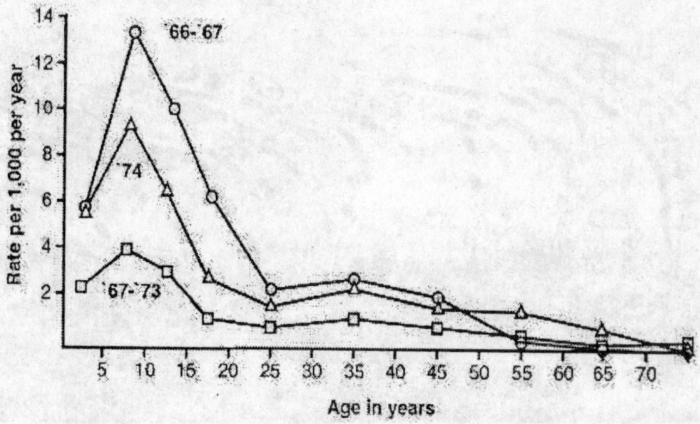

Fig. 3.14. Incidence of M. pneumoniae pneuminia in Seattle by age, for two epidemics (1966-67 and 1974) and the endemic periods (1967-73).

youngest age group, 0 to 4 years old, were about one-half those in school-age children, but considerably higher than in adults. Mycoplasma pneumoniae pneumonia was rarely observed in infants younger than 6 months, suggesting maternally conferred immunity. *Mycoplasma pneumoniae* accounts for 8 to 15 percent of all pneumonias in young school-age children. In older children and in young adults, the organism is responsible for approximately 15 to 50 percent of all pneumonias. Infection with *M. pneumoniae* occurs worldwide all year round but shows a predilection for the colder months, apparently because of the greater opportunity for transmission by droplet infection. *Mycoplasma pneumoniae* appears to require close personal contact to spread; successful spreading usually occurs in families, schools, and institutions. The incubation period ranges from 2 to 3 weeks.

Ecology

There are no free-living mycoplasma that are not parasites. They parasitize a wide range of organism including humans, plants, animals, and insects. mycoplasma grow very slowly, even under perfect conditions, with a generation time ranging up to nine hours in some species. They also have a very long lag phase, so it may take an entire week before colonies become visible on agar plates. Due to their degraded genome, and inability to perform basic functions, mycoplasma rely on their host for much of their nutrition.

Fig. 3.15. Mycoplasma hyopneumoniae attached to swine cilia.

Many species of mycoplasma are commensal, living innocuously with their host as part of the natural flora of the body, and having no detrimental effects. But when mycoplasma act as pathogens they cause chronic, but mild infections, rarely, if ever, killing their host. They are usually surface parasites, although several species, including the aptly named *M. penetrans*, are intracellular pathogens.

Laboratory Diagnosis

The laboratory diagnosis of mycoplasma infection can be accomplished by:

1. Culturing the organism from sputum, mucous membrane swabbings or other specimens by direct inoculation into liquid or solid media containing serum, yeast extract and penicillin to inhibit contaminating bacteria. Colonieswill become detectable in one to three weeks. They stain intensely with neutral red or tetrazolium or methylene blue. The organism can be presumptively identified by its hemabsorption or B-hemolysis of guinea pig red blood cells. Staining its colonies with homologous fluorescein-labelled antibody conclusively identifies it. A routine mycoplasma medium consists of heart infusion, peptone, yeast extract, salts, glucose or arginine, and horse serum (5 to 20 percent). Fetal or newborn calf serum is preferable to horse serum. To prevent the overgrowth of the fast-growing bacteria that usually accompany mycoplasmas in clinical materials, penicillin, thallium acetate or both are added as selective agents. For *Ureaplasma* culture, the medium is supplemented with urea and its pH is brought to 6.0. *Ureaplasma* and *M. genitalium* are relatively sensitive to thallium, which is, therefore, omitted from their culture media. For *M. pneumoniae* isolation, nasopharyngeal secretions are inoculated into a selective diphasic medium (pH 7.8) made of mycoplasma broth and agar and supplemented with glucose and phenol red. When *M. pneumoniae* grows in this medium, it produces acid, causing the color of the medium to change from purple to yellow. Broth from the diphasic medium is subcultured to mycoplasma agar when a color change occurs, or at weekly intervals for a minimum of 8 weeks.
2. Quantitation of the patient antibody response to mycoplasma by complement fixation tests on acute and convalescent

serum. Cold agglutinins to human Orythrocytes may also be measured. Colonies appearing on the plates can be identified as *M. pneumoniae* by staining directly on agar with homologous fluorescein-conjugated antibody or by demonstrating that a specific antiserum to *M. pneumoniae* inhibits their growth on agar. Colonies of ureaplasmas are usually minute (less than 100 µm in diameter); because of urea hydrolysis and ammonia liberation, the medium becomes alkaline. When manganous sulfate is added to the medium, the ureaplasma colonies stain dark brown. Isolates can be characterized in more detail by a variety of biochemical and serologic tests. More sophisticated tests, including electrophoretic analysis of cell proteins, DNA-DNA hybridization tests, mycoplasmal DNA cleavage patterns by restriction endonucleases, and PCR tests employing species-specific primers for amplification, may be performed in a research laboratory.

3. Serodiagnosis consists of examining serum samples for antibodies that inhibit the growth and metabolism of the organism or fix complement with mycoplasmal antigens. A fourfold or greater rise in IgG titer is considered indicative of recent infection, whereas a sustained high antibody titer may not be significant, because a relatively high level of antibody may persist for at least 1 year after infection. A variety of rapid tests based on indirect hemagglutination of erythrocytes or latex particles coated with *M. pneumoniae* antigens have been developed, and some are commercially available.The cold agglutinin test is less useful because only about one-half of patients develop cold agglutinins and because these antibodies also are induced by a great many other conditions.

Present techniques for laboratory diagnosis of *M. pneumoniae* infections are of little use to the clinician because recovery by culture and identification of the mycoplasmas take at least 1 to 2 weeks. Methods for rapid laboratory diagnosis, such as direct demonstration of organisms in the respiratory specimens by nucleic acid amplification techniques, have promise but diagnostic kits are not yet commercially available.

There are five basic steps when setting up a good mycoplasmal testing program:

1. Test all of cell lines that are currently in use to eliminate these cultures as potential sources of contamination.

2. Quarantine and then test all cell lines coming into the lab from any outside sources.
3. Test continuously maintained cell cultures at defined intervals (most experts recommend monthly to quarterly).
4. Make sure all new additions to cell repository are thoroughly tested at the time of freezing. This saves both time and money by eliminating the needs for testing these cultures each time they are thawed.
5. Test new lots of sera whenever possible, and always buy from a reputable source that tests each lot. If serum testing is not feasible, be sure to carefully retest cells after they have been cultured for 4-6 weeks in media containing the serum.
6. Antibodies (IgG, IgM and IgA) against peptide-specific mycoplasma should be performed simultaneously.
7. As no diagnostic tool is 100% accurate, suggest that PCR, molecular probe, and IgG, IgM, and IgA antibodies should all be performed to gain the most accurate result.

Importance of Measuring IgG and IgM Antibodies against Mycoplasma fermentans

It has developed a specific ELISA assay for measurement of antibodies against mycoplasma fermentans and compared the results to the presence of DNA in the blood. It has found that only in about 60% of cases where *M. fermentans* was positive, antibodies to *M. fermentans* antigens were elevated significantly. In the other 40% in which the genome was positive, IgG and IgM antibodies were not detected. This may be due to the nature of the *M. fermentans* cell invasion, the inhibition of phagocytosis, and the lack of immune response to this organism in these individuals. On the contrary, in about 20% of cases, the *M. fermentans* genome was absent but antibodies of IgG and IgM isotype were detected in their blood. The absence of *M. fermentans* DNA from blood cells and the simultaneous presence of antibodies to this mycoplasma in the serum of the same patients

suggest chronic infection of other tissues or cells with *Mycoplasma fermentans*. Another possibility is that these antibodies are cross-reactive in their nature. This means that antibodies produced against collagen, cartilage, and thyroid in some patients with autoimmune disease may cross-react with mycoplasma antigens and give false positive results. For this reason, it is measured antibodies against synthesized peptides corresponding to *M. fermentans* and were able to reduce the degree of cross-reactivity.

CABRI collections have evaluated and compared the usefulness, practicality and costs of a number of different methods for their daily application in the routine setting of a cell culture bank. It is agreed that a combination of various tests is the most useful approach for the routine detection of these contaminats, and that the most reliable assays are the following: (direct DNA fluorescence staining, classical broth-agar microbiological colony assay, RNA hybridization in solution and PCR). In addition, other assays can be used, which are available in kits, like the Mycotect system (which requires co-cultivation of cells to be tested with 6-methilpurine deoxyriboside (6-MPDR)), the Mycoplasma Detection Kit (an enzyme immunoassay), etc. It has to be kept in mind that these assays present both advantages and disadvantages, and that in no case a single assay is sufficient for exclude the contamination of a culture.

In selected cases, an attempt is made to eliminate mycoplasma contamination by treatment with specific antibiotics: MRA (Mycoplasma Removal Agent) containing a quinolone derivative, Ciprobay containing ciprofloxacin, Baytril containing enrofloxacin, or BM Cyclin containing tiamulin and minocycline.

Control and Prevention

Chemoprophylaxis of mycoplasma infections is not recommended, and no vaccine is available. Prior natural infection appears to provide the most effective resistance; however, evidence shows that *M. pneumoniae* infections recur at intervals of several years. These observations suggest that immunity to a single natural infection is relatively short-term. Primary atypical pneumonia is usually selflimiting and does not require antibiotic treatment. However, if antibiotics are needed, the drug of choice is one of the macrolide antibiotics:

Azithromycin
Clarithromycin
Dirithromycin
Erythromycin

Urogenital diseases may be treated with:

Metronidazole (except during the first trimester of pregnancy)
Clindamycin

Treatment

The mycoplasmas are sensitive to tetracyclines, macrolides, and the newer quinolones, but are resistant to antibiotics that specifically inhibit bacterial cell wall synthesis. Tetracycline or erythromycin is recommended for treatment of *M. pneumoniae* pneumonia, although effective treatment of the symptoms usually is not accompanied by eradication of the organism from the infected host. To prevent recurrence of nongonococcal urethritis caused by *U. urealyticum*, sexual partners should be treated simultaneously with tetracycline. The incidence of tetracycline-resistant strains of *U. urealyticum* and *M. hominis* is on the rise.

Mycoplasmas – Stealth Pathogens

While *Mycoplasma pneumonia* is certainly not the only species causing disease in humans, it makes for a good example of how this stealth pathogen can move out of it's typical environment and into other parts of the body and begin causing other diseases. While residing in the respiratory tract and lungs, *Mycoplasma pneumonia* remains an important cause of pneumonia and other airway disorders, such as tracheobronchitis, pharyngitis and asthma. When this stealth pathogen hitches a ride to other parts of the body, it is associated with non-pulmonary manifestations, such as blood, skin, joint, central nervous system, liver, pancreas, and cardiovascular syndromes and disorders. Over the past 20 years the annual number of reports on extrapulmonary symptoms during *Mycoplasma pneumoniae* disease has increased. Clinical and epidemiological data indicate that symptoms from the skin and mucous membranes, from the central nervous system, from the heart, and perhaps from other organs as well are not quite uncommon manifestations of *M. pneumoniae* disease.

This single stealth pathogen has been discovered in the urogenital tract of patients suffering from inflammatory pelvic disease, urethritis, and other urinary tract diseases. It has been discovered in the heart tissues and fluid of patients suffering from cardititis, pericarditis, tachycardia, hemolytic anemia, and other coronary heart diseases. It has been found in the cerebrospinal fluid of patients with meningitis and encephalitis, seizures, ALS, Alzheimer's and other central nervous system infections, diseases and disorders. It has even been found regularly in the bone marrow of children with leukemia. It is amazing that one single tiny bacterium can be the cause of so many seemingly unrelated diseases in humans. But as with all mycoplasma species, the disease is directly related to where the mycoplasma resides in the body and which cells in the body it attaches to or invades.

Method for Detecting Mycoplamas

Unlike contamination caused by most bacteria, yeast or fungi, mycoplasmas typically do not alter the turbidity or pH of cell culture media. Thus, they can only be detected through an active effort employing highly specialized and specific tests that require a skilled operator to perform and interpret results. Tests fall into two basic categories: direct culture with media; or indirect tests measuring specific biochemical markers or other characteristics associated with mycoplasmas.

Direct culture requires the use of one or more complex nutritionally enriched mycoplasmal media and carefully controlled environmental conditions. Even then, some mycoplasmal strains are difficult to grow in culture without cells. Properly done, with appropriate positive and negative controls, direct culture testing offers the greatest security, but is rather slow, usually requiring up to 28 days for completion.

There are many indirect tests of varying sensitivity and convenience, including DNA fluorochrome staining, DNA probes, PCR, ELISA, autoradiography, immunofluorescence and specific biochemical assays. While faster than direct culture methods, indirect tests are not yet as sensitive and usually require higher levels of contamination (i.e. 104+ organisms/mL) for detection. Indirect tests offer two advantages over direct culture methods: first, they can detect so-called "non-cultivable" mycoplasmal strains that direct culture may miss; second, they are faster, usually taking only 1-5 days to complete.

4

PHYTOPLASMA

When phytoplasmas were first discovered in plants, they were called mycoplasma-like organisms because of their similarity in appearance to mycoplasmas using transmission electron microscopy. In 1994, an article was published in the International Journal of Systematic Bacteriology that changed the trivial name to phytoplasma. (Seemuller et. al. 1994). The Phytoplasma or mycoplasma-like organisms (MLOs) are a unique and intractable group of plant pathogens. They cause economically important diseases in a broad range of annual and perennial host plants. Phytoplasma are cell wall-less prokaryotes (class Mollicutes) which infect plants. They live in the phloem, are transmitted by grafting, dodder, and insects, and cause symptoms such as yellowing, phyllody, and witches' brooms. It is believed that phytoplasmas cause symptoms of yellowing, phyllody, and witches' brooms due to an imbalance in the hormone content of the plant. Hormones are very hard to measure in plants because they are present in such tiny amounts, so this theory has never been proven. Before it was discovered the disease aster yellows was caused by a phytoplasma, scientists did not know these strange organisms existed. They thought a non-mechanically transmissible virus caused aster yellows disease. Viral particles could not be identified in aster yellows infected plants, although the disease was dodder, graft, and insect transmissible. Insects like leafhoppers, planthoppers, and psyllids transmit Phytoplasma. These are the most common vectors of phytoplasma. There is an element of host specificity on the part of the phytoplasma and on the part of the leafhopper. Some leafhoppers only feed on particular plant species, so this is a limitation. Phytoplasmas (former mycoplasma-like organisms) are non-cultivable plantpathogens of the class Mollicutes. They are restricted to phloem sieve elements, where they are generally present in low concentrations, especially in woody plants. During the past decade, research has yielded new knowledge about the plant and insect host ranges, geographical distribution, and phylogenetic relationships of phytoplasmas, and a taxonomic system has emerged in which distinct phytoplasmas are named as separate "Candidatus phytoplasma species." In large part, this progress has resulted from the development and use of molecular methods to detect, identify, and classify phytoplasmas. While these advances continue, research has recently begun on the phytoplasma genome, how phytoplasmas cause disease, the role of mixed phytoplasmal infections in plant diseases, and molecular/ genetic phenomena that underlie symptom development in plants. These and other recent advances are laying the foundation for future progress in understanding the mechanisms of phytoplasma pathogenicity, organization of the phytoplasma genome, evolution of new phytoplasma strains and emergence of new diseases, bases of insect transmissibility and specificity of transmission, and plant gene expression in response to phytoplasmal infection, as well as the design of novel approaches to achieve effective control of phytoplasmal diseases. Mycoplasma-like organisms (MLOs) are found in plants. They differ from mycoplasmas in that MLOs are intracellular (inside the host cells). Most MLOs cause yellowing or stunting of the host, but a few induce extra growth. MLOs are transmitted between plants by vectors such as plant-hoppers and psyllids. MLOs are associated with (1) gross changes in flower structure, (2) the development of leaf-like structures in plant inflorescences (phyllody), (3) leaf proliferation, and also with (4) some witches' brooms. For example: clover phyllody, potato witch's broom.

General Characteristics

All efforts at charatcterization have been frustated by the inability of researchers to culture MLOs axenically. Of necessity, therefore, research on these agents has been conducted in either the host plants in which they cause disease or in the vector insects (usually leafhoppers), which transmit them. In practice, this has meant that normally trivial procedures such as the inoculation of plants with the pathogen, the preservation of voucher specimens and the comparison of related agents pose challenging technical difficulties (McCoy, 1989). Latent period of weeks to months is required for the organisms in both their plant and insect hosts. Detection techniques are often laborious to perform, requiring time-consuming bioassays or electron microscopy, although in recent years promising molecular detection techniques have been developed for some MLOs (Kirkpatrick et al., 1987; Lin and Chen, 1985). While graft transmission of the agents between related plants is possible, insect transmissibility is quickly lost when MLOs are maintained by grafting. Therefore, research on these pathogens in a transmissible form requires the maintenance of an elaborate biosystem, which includes leafhoppers, plants and the phytoplasma. Phytoplasmas may best be described as bacteria that have lost their cell walls. They are only enclosed by a thin membrane which makes them very pliable and able to assume a variety of shapes (think of a balloon filled with water that can be squeezed into different shapes).

Biological Properties

An obligately parasitic habit, structural fragility, presence in low numbers in infected plants, and intimate association with host tissues are all attributes of phytoplasmas that previously hindered their characterization and precluded their definitive classification. The first attempts to differentiate and classify phytoplasmas relied upon knowledge of certain biological properties such as symptomatology, the range of plant hosts affected and specificity of transmission by vector insects suggested that phytoplasmas could be broadly classified into three main groups namely, decline, proliferation and virescence agents according to the types of symptoms that these agents induced on their respective plant hosts. classified phytoplasmas inciting numerous yellows diseases in Japan into three strain groups, based on the transmissions characteristics of these pathogens by the leafhopper *Macrosteles orientalis* to a range of plant species, and by the types of symptoms that developed on infected plants. Similarily, differentiated a wide range of phytoplasma diseases into two major categories according to either the mutually exclusive floral symptoms of virescence and phyllody or reduced flower size and color that the associated pathogens induced on the experimental host periwinkle (*Catharanthus roseus*).

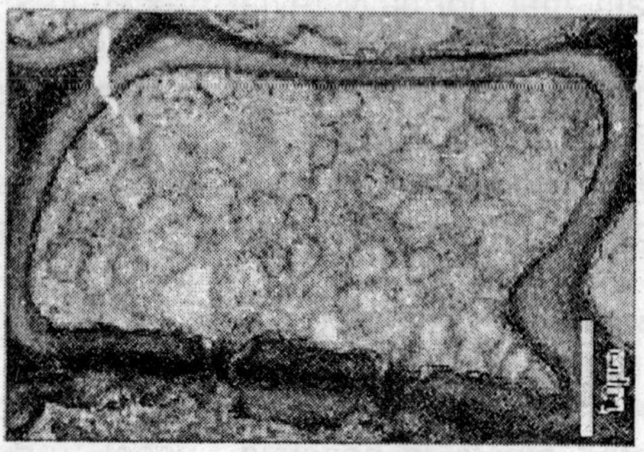

Fig 4.1. Transmission electron micrograph of a ultrathin sectionof the sieve element of the leaf midrib snowing a large number of phytoplasmas.

Molecular Properties

There is little information available on the genetic architecture of the non-helical, phytopathogenic mollicutes, the phytoplasmas. The G+C content of several phytoplasmas is at the lower end of the spectrum for mollicutes (Kollar and Seemüller, 1989; Sears et al., 1989). Phytoplasma genomes have been shown to contain two 16S rRNA genes (Schneider and Seemüller, 1994), ribosomal protein genes (Lim and Sears, 1992), a tuf gene encoding the elongation factor EF-Tu (Schneider et al., 1997), and genes encoding a major membrane protein gene (Barbara et al., 1998), an antigenic protein (Yu et al., 1998) and a nitroreductase (Jarausch et al., 1994). Extrachromosomal DNA has been reported for some phytoplasmas (Davis et al., 1988; Sears et al., 1989; Davis et al., 1990; Kuske and Kirkpatrick, 1990; Kuboyama et al., 1998), although its role remains speculative. The genome sizes of several phytoplasmas have been also determined and its range is from 530 to 1185 kb (Neimark and Kirkpatrick, 1993; Zreik et al., 1995; Firrao et al., 1996a, b; Marcone et al., 1999). Little else is known about the genetic structure of phytoplasmas, and genes involved with pathogenicity and other general housekeeping genes have not been identified.

Serological Properties

The introduction of molecular methods to studies on phytoplasma diseases was facilitated by significant improvements in phytoplasma extraction from infected plant and insect hosts. In some instances, phytoplasma-enriched extracts were used as a source of immunogens to produce polyclonal or monoclonal antibodies against several phytoplasmas. This facilitated development of serological assays for pathogen detection and identification. However, difficulties encountered in generating specific high titer polyclonal antisera restricted their production mostly to phytoplasmas maintained in high titer herbaceous plant hosts such as periwinkle. By comparison, monoclonal antibodies have demonstrated considerable promise for serological detection identification of phytoplasmas from a much broader range of host plants, including woody perennials. Exhibiting a high degree of specificity, monoclonals seem to be most suited for differentiating very closely related strains.

Phytoplasma-enriched preparations were also utilized as a source of DNA for isolation and cloning. The first estimates of phytoplasma DNA composition were reported in for DNAs enriched by repeated fractionation on cesium chloride-bisbenzimide density gradients. The low values (23 to 26.3 mol %) reported for G+C content were within the known range of the cultivable Mollicutes, as were later size determinations for linearized phytoplasma genomes (640-1185 kb) resolved by pulse field gel electrophoresis.

Cloned fragments of phytoplasma DNA have been widely employed as probes in dot hybridizations assays to detect and identify phytoplasmas. Used for Southern hybridization in conjunction with restriction fragment length polymorphism (RFLP) analysis, probes enabled investigations of genetic relationships among phytoplasmas associated with similar hosts or with symptomatologically similar diseases. This type for analysis comprising strains that shared extensive sequence homology and little or no apparent homology with other phytoplasmas, identified several discrete phytoplasma groups, each coined the term genomic strain cluster to denote each of seven discrete genotypic groups, resolved by employing a selection of phytoplasma genomic DNA probes.

Serological techniques can also be used to identify protein antigens associated with phytoplasmas that may be important in how they function as pathogens. Polyclonal and monoclonal antibodies to aster yellows isolates belonging to Group 1B recognise a single immunodominant protein (MMP) located on the membrane surface that appears to be common to all phytoplasma isolates in the same sub-group. This paper describes the cloning and sequencing of the gene determining the MMP for the chlorantie isolate of aster yellows, and the subsequent cloning of a gene determining an homologous protein from clover phyllody, a serologically unrelated phytoplasma from a different subgroup of aster yellows.

The use of serology has yielded results that are complementary to those obtained by molecular means but without the sensitivity of PCR. The predicted properties of the translation product suggest a membrane protein with most of the mature protein displayed outside the cell. An N-terminal export sequence of 32 residues and cleavage point would leave a mature protein of 201 residues (which is very close to the size expected from electrophoretic studies of the protein purified from infected plants) anchored in the cell membrane by a trans-membrane domain located close to the C-terminal. A short C-terminal sequence remains inside the cell. This protein was identified as the membrane protein for two reasons. Firstly, there was complete agreement between the amino acid sequence of the translation product and the original amino acid sequences derived from the purified membrane protein preparations at the beginning of the work. This included the original amino acid sequences derived from the N-terminal of the mature membrane protein molecule, which matched those of the N-terminal of the mature protein after cleavage of the export domain. Secondly, in preliminary studies on expressing the gene in *E. coli* there was serological reaction with monoclonal antibodies raised against the native protein.

Several primer pairs were also used with DNA prepared from a range of other phytoplasma isolates. Most primers examined yielded a product only from other phytoplasmas belonging to the same aster yellows subgroup as chlorantie. One primer pair designed to sequences before and after the membrane protein gene yielded a product from an isolate of clover phyllody. The product was smaller than that obtained from the aster yellows isolates and contained an ORF of 495 bases with a putative translation product of 164 residues. There was a high degree of sequence similarity between the two PCR products except for a region in the ORF that corresponded to the hydrophilic region of the MMP. The clover phyllody protein contained a similar N-terminal export sequence to the aster yellows membrane protein, also 32 residues long with a sequence similarity of 87.5%. There was greater than 86% sequence similarity at the C-terminus including a trans-membrane domain with more than 70% homology. However, there was only 17% to 28% sequence similarity between the two proteins for the major hydrophilic domain predicted to be displayed on the surface of the cell.

It is likely that the two proteins are homologous in function as indicated by their similar overall structure. However, the function or functions are still unknown. Attempts to obtain stable expression of the protein in conventional bacterial expression systems have met with difficulties, as the protein appears to be toxic to the bacterial cell. The relatively low sequence similarity in the hydrophilic domains predicted to be displayed on the membrane surface probably reflects how the different pathogens may have adapted to their respective host plants or insect vectors. This is also consistent with the lack of serological cross-reactivity between the two phytoplasmas, which are related taxonomically in that they both fall into the aster yellows group, but in different sub-groups. The membrane proteins from both aster yellows and clover phyllody phytoplasmas are different in amino acid sequence and overall structure from a membrane protein associated with the apple proliferation phytoplasma and a protein antigen associate with sweet potato witches broom, both of which belong to unrelated taxonomic groups.

Structure

One of the major difficulties of working with phytoplasma is the inability to grow the organism in culture. As long as phytoplasmas are not cultured in vitro, and even though they retain their viability for several hours in phytoplasma-enriched extracts from infected plant or insect vectors, genetic manipulation of these organisms is still very speculative. Due to their high sensitivity, polymerase chain reaction methods are thus well suited to the detection and identification of phytoplasmas.

Structural Genomics

Only limited efforts have been directed toward cloning and sequencing phytoplasma genes. Their genomes consist of a circular, 640-1200 kb chromosome of low G+C content. Phytoplasmas extrachromosomal DNAs revealed surprising sequence similarity with both bacterial plasmid of gram-positive bacteria and plant DNA viruses. The evidence of intermolecular recombination between extrachromosomal DNAs is suggestive of the origin of those peculiar chimeric molecules. For the phytoplasma causing apple proliferation disease (AP phytoplasma), a few genes could be identified and a physical map of the 645 kb chromosome was recently constructed from several gene sequences. Classification and determination of the phylogenetic position of phytoplasmas is mainly based on 16S rRNA and ribosomal protein sequences, leading to the establishment of one monophyletic clade consisting of 14 major groups. The 16S rRNA gene was targeted in an attempt to obtain sequence data for determining the evolutionary origin of phytoplasmas. Full-length 16S rRNA genes of phytoplasmas associated with virescence of evening primrose, western X and western aster yellows disease were cloned and sequenced by conventional methods. Resulting phylogenetic analyses of gene sequences clearly established phytoplasmas as members of the class Mollicutes, while also showing them to be more closely related to one another than to any of the cultivable Mollicutes. Their unique status within the Mollicutes has since been corroborated by phylogenetic analyses of additional, less-conserved, evolutionary markers; the ribosomal protein genes rps3 and rpl2 the 16-23S rRNA spacer region and the gene encoding elongation factor Tu (tuf gene).

The polymerase chain reaction (PCR) incorporating mollicute-specific oligonucleotide primers derived from rRNA sequences made selective amplification of near full-length phytoplasma 16S rRNA genes from mixtures with host plant DNA possible. By its simplicity of application and superior sensitivity, PCR quickly became established as the method of choice for detection and diagnosis of phytoplasma diseases. Phytoplasma-specific rRNA primer sets permitting amplification of various regions of one or both rRNA operons from some or all known phytoplasmas. Collective profiles revealed by separate digestion of rDNA products with a selection of frequently cutting endonuclease enzymes varied, according to the particular phytoplasmas that were examined. Strains considerd to be closely related due to similar or shared biological properties were often, although not always, indistinguishable on the basis of RFLP patterns, whereas pairwise comparisons of disparate strains were marked by considerable differences in RFLP patterns. Furthermore, cluster analysis of similarity coefficients derived from rDNA RFLP patterns provided for the first time a means to differentiate and classify both known and unknown phytoplasmas from a wide range of plant hosts and geographic locations into well-defined groups and subgroups of related strains. The spacer region (SR) separating the 16S from the 23S rRNA gene of phytoplasmas was also shown to be a reliable phylogenetic marker. Phylogenetic trees derived from the entire 16-23S SR or variable regions flanking the tRNAile gene differentiated phytoplasmas into groups concordant with the major groups previously established from analyses of 16S rRNA genes. Several benefits to using 16-23S SR sequences for the purpose of phytoplasma classification were also realized. Since the total length of the sequence is about 250 bp, a full-length sequence through this region can be readily obtained in both directions with only two sequencing primers. Direct sequencing from PCR products is often easier because of higher yields with smaller PCR products and the fact that sequences are inherently more variable than 16S rRNA genes. Phytoplasmas differ in their 16S rDNA sequence by no more than 14%, whereas their respective 16-23S SR differed by as much as 22%.

Phytoplasma RNase P RNA deviates from the bacterial consensus and reveals novel structural elements

The primary sequence of RNase P RNA from the aetiological agent of apple proliferation disease (AP phytoplasma, strain AT) is most similar to those from the Mycoplasma cluster, including *Acholeplasma laidlawii*; the extremely low G-C content of 26.4% is consistent with the biased base composition typical for phytoplasma genomes. The postulated secondary structure corresponds to the type B RNase P RNAs. One major difference between the ancestral (A) and Bacillus-type (B) classes is the lack of the distal helices P16 and P17 in the B-type structures and, as a consequence, of the long-distance pairing P6 between L17 and J5/7. A comparable interaction between L5.1 and L15.1 has been implied from covariation analysis of type B sequences; this helix (P22) does not exceed two consecutive base pairs in any of the known RNase P RNAs. In striking contrast, the postulated P22 in the phytoplasma RNA consists of 5 bp (or up to 8, if P5.1 opens at the distal end), and thus strongly supports the existence of this interaction in all type B RNase P RNAs. The unusual length of P22 may be required to stabilize the tertiary structure of this extremely A-U rich ribozyme. The unambiguous confirmation of existence and function of this interaction, however, will require additional sequence data and corresponding functional studies.

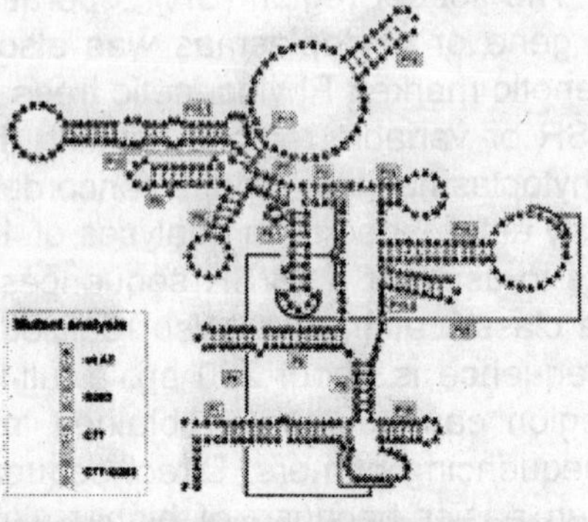

Fig 4.2. Secondary structure of phytoplasma RNase P RNA.

Another B-type peculiarity is the extended P10.1 close to the cruciform region. A postulated tertiary interaction between a receptor motif in P10.1 and the L12 tetraloop may result in overall stabilization of the ribozyme structure; this view is supported by the observation that RNase P RNAs lacking P12 are catalytically active under conditions of increased ionic strength. L12 is GAAA in all B-type RNase P RNAs containing the P10.1 extension, but UAAA in the phytoplasma RNA. The existence of a similar tertiary interaction despite the sequence divergence is supported by covariation of the length of P12 and P10.1.

The phytoplasma RNase P RNA deviates at two additional positions from the established bacterial ribozyme consensus: a U-A pair in this organism replaces the fully conserved proximal C-G pair of P5. The colinearity of genomic and RNA-derived sequence (obtained by RT-PCR) excludes any post-transcriptional sequence changes, such as RNA editing, during expression of this RNA. As for the L12 sequence, these changes might thus be explained by base transitions that have occurred to accommodate the sequence to the low G-C content in the phytoplasma genome.

A consensus base pair close to the catalytic center modulates, but is not essential for ribozyme activity of phytoplasma RNase P RNA

Sequence conservation and functional data have led to the conclusion that the catalytic core of bacterial RNase P RNAs is constituted at least in part by the conserved helical element P4. The absolute conservation of the adjacent C-G pair suggests a significant role in the function of bacterial RNase P RNAs. The unexpected discovery of a U-A pair at the corresponding position in the phytoplasma RNA implies, however, that the conserved sequence may not be important for RNase P function in vivo. It has been thus investigated the influence of nucleotide identity and base-pairing properties at this position on ribozyme function by converting the U-A back to the consensus C-G, and by introducing a 'wobble' U·G or a non-canonical C·A pair, respectively.

The steady-state parameters, Kmapp and kcatapp of the wild-type phytoplasma ribozyme are 116 nM and 0.75 min-1, respectively, and are well within the range described for other bacterial RNase P RNAs. Remarkably, although these parameters are significantly different for the (A+U)-rich RNase P RNA from *Mycoplasma hyopneumoniae*, the specificity constant kcatapp/Kmapp of that ribozyme is close to that observed in the phytoplasma RNA. The disruption of the Watson-Crick U-A pair, or the conversion to the consensus C-G, results in moderate to significant changes in the catalytic properties of the phytoplasma ribozyme if tested under the conditions established for the wild-type RNA for the mutants lacking the Watson-Crick pair, Kmapp increases to >2-fold, indicating an impairment in substrate binding. Concomitantly, kcatapp increases in the U·G and C·A variants. This might be explained by the observation that moderately reduced affinities can result in higher substrate turnover numbers. In contrast, kcatapp remains essentially unchanged for the C·G mutant. Consequently, all three mutants show a reduction in catalytic efficiency compared to the wild-type due to their increase in Kmapp.

It has been shown previously that the effect of mutations or deletions at conserved positions in RNase P RNA can frequently be overcome by increased ionic strength. Thus the kinetic parameters determined under the ionic conditions needed for optimum activity of each RNA mutant. All phytoplasma RNase P RNA variants require 2 M NH_4Cl for maximum activity. In contrast, the optimal concentrations of $MgCl_2$ are higher for the mutants containing non-Watson-Crick base pairs, than for the wild-type and C-G mutants, respectively. The requirements for high mono- and divalent cation concentrations thus resemble those determined for *Bacillus subtilis* RNase P RNA) and are unlike those of *E.coli* M1 RNA, which belongs to the structurally different A-type. Such a high ionic strength environment may help to alleviate minor deviations from an optimal RNase P RNA structure by supporting the correct folding of the ribozyme core. This hypothesis is supported by the finding that, under increased Mg^{2+} concentrations, a decrease of Kmapp is observed for the two 'non-Watson-Crick' mutants, which approaches wild-type level and is paralleled by a significant drop in kcatapp for the U·G mutant. Thus, the findings of this experiment support the hypothesis that, although the conserved C-G pair may modulate catalytic efficiency of bacterial RNase P RNAs, neither its presence as a Watson-Crick pair nor its sequence alone are crucial for efficient ribozyme function. Interestingly, two deviations from the bacterial consensus at positions different from those in phytoplasma are present in RNase P RNA from a primitive plastid, which is not a ribozyme in vitro. Restoration of the consensus does not confer ribozyme activity on this RNA; however, heterologous reconstitution with a cyanobacterial protein subunit yields a functional holoenzyme, irrespective of the RNA sequence. Taken together, these results support the notion that even fully conserved nucleotides in the central region of *B. subtilis* RNase P RNA might not be essential for catalysis, but rather for stabilization of a favorable tertiary structure. The recent finding that the conserved helix P4 is a metal binding domain, possibly serving a structural role in the folding of the ribozyme core, corroborates this view. The Mg^{2+}-binding properties of P4 may well be influenced by adjacent structural elements. Accordingly, the requirement for higher concentrations of Mg^{2+} in the phytoplasma RNase P RNA mutants containing non-Watson-Crick pairs suggests that the terminal base pair could play a role in stabilization of P5 or other structural elements adjacent to the catalytic core, and-among other possible reasons-may influence the binding of functionally important divalent cations.

The first sequencing strategy based on *S. citri* chromosome map, led to the acquisition of 5,500 reads giving an assembly covering 24 % of the 1.8 Mb *S. citri* chromosome. Since this approach did not lead to completion, three new libraries were constructed. Two shotgun plasmid libraries of 4,000 and 2,500 clones were constructed, with mechanically sheared inserts, with sizes of 3.4 and 1.5 kb, respectively. In order to facilitate the medium distance assembly, a miniBAC library containing 768 clones with insert of 20 kb was also built up, after Sau3a partial digestion of *S. citri* chromosome. Up to now, about 13,000 reads have been acquired and assembled into nearly 500 contigs covering 71 % of the *S. citri* chromosome.

Functional Genomics

Fructose operon mutants of *S. citri* have been obtained both by transposition (mutant GMT553) and through homologous recombination. These mutants that are unable to use fructose were found to be much less pathogenic than the wild-type strain. When mutant GMT553 was complemented with the fructose operon genes that restored fructose utilization severe pathogenicity similar to that of the wild-type strain was also restored. Fructose utilization by the spiroplasmas was thought to impair sucrose loading into the sieve tubes by the companion cells leading to accumulation of carbohydrates in source leaves and depletion of carbon sources in sink tissues. The first gene fruR of the fructose operon has been characterized as an activator of the fructose operon.

New oriC plasmid vectors have been constructed to improve gene disruption through homologous recombination in *S. citri*. These vectors have been successfully used to target *S. citri* genes such as the motility gene scm1, mreB encoding a cytoskeleton protein, and spiralin the major surface protein. The corresponding mutants are currently being characterized.

Through the sequencing of ribosomal DNA new phytoplasmas have been characterized. A phytoplasma inducing a lethal disease in almond trees has been fully characterized by sequencing its 16S rDNA. It is a new phytoplasma in the PPWB / 16SrIX-A group. A PPWB group phytoplasma is also found in Prunus. Phytoplasmas belonging to the PPWB group, but different from the almond phytoplasma were also found in naturally infected Lactuca serratia and periwinkle. Tomato and pepper plants were found infected by new phytoplasmas belonging to the CP group.

In addition to ribosomal sequences, tuf gene sequences proved to provide relevant taxonomic information and have been used to better understand the taxonomic structure of the Aster Yellows phytoplasma clade.

Different procedures were used to compare FD isolates from France, Italy and other phytoplasma in the EY group from grapevine (PGY, Germany) or from other hosts (elm, alder, hemp dogbane). RFLP analyses, HMA and sequence data from two

DNA fragments, one ribosomal and one extra ribosomal (FD9) were used to construct dendrograms or neighbor-joining trees. All data showed that FD sensu stricto (i.e., grapevine phytoplasma in the EY group transmitted by the leafhopper vector *S. titanus*) all clustered in close relationship separately from other EY phytoplasmas (PGY not transmitted by *S. titanus* and isolates EY1, HD1, ULW). However, ALY, isolated from alder in Italy, always clustered in the same subgroup as FD sensu stricto. Some epidemiological traits might be reflected by such differences in isolates.

PCR-RFLP of a non ribosomal fragment of apple proliferation phytoplasmas has revealed variability. Three subtypes (AP, AT-1 and AT-2) were evidenced and characterized. They differ only by point mutations in the restriction sites of enzymes used in RFLP. No geographical of a given subtype could be observed. Thus phytoplasmas infecting *Malus* sp. and associated with apple proliferation disease are genetically homogenous. Similar studies were carried out with phytoplasmas of the AP group infecting *Prunus* sp. and designated under the name ESFY. This showed that the ESFY phytoplasmas were genetically homogenous but different from those infecting *Malus* sp. Therefore they are likely to represent different taxa. ESFY phytoplasmas were detected in 6 wild *Prunus* sp. and in *Rosa canina*, *Fraxinus excelsior* and *Celtis australis*.

At 622 kb, the SPLL-V4 chromosome is one of the smallest of the phytoplasmas with only two Bermuda grass white leaf (BGWL) isolates having a smaller chromosome size of 530 kb (Marcone et al., 1999). The size of the TBB chromosome was estimated to be 662 kb, which is smaller than the 690 kb estimated by Marcone et al. (1999) from full-length chromosomes. This discrepancy could be due to different electrophoresis conditions, anomalies in the separation of larger compared to smaller, digested fragments, or too different isolates being used. Within the faba bean phyllody (FBP) phytoplasma group to which strains SPLL-V4 and TBB belong, chromosome sizes fall within a narrow range and include 660 kb for the faba bean phyllody phytoplasma (Marcone et al., 1999), 720 kb for witches' broom disease of lime (Zreik et al., 1995) and 790 kb for the phytoplasma associated with cotton phyllody (Marcone et al., 1999). The size of the SPLL-V4 phytoplasma genome is the smallest within this group. Genomes of the aster yellows (AY) phytoplasmas were found to be much larger (Neimark and Kirkpatrick, 1993), but recent genome size estimations of 29 AY isolates have shown that chromosome sizes within this group are heterogeneous and range from 660 kb to 1130 kb (Marcone et al., 1999). Interestingly, phytoplasmas in the AY group are widespread and have been reported in several continents, whereas phytoplasmas in the FBP group have only been identified in Australia and Asia (Seemüller et al., 1998). As more phytoplasma genomes are studied, a relationship between geographic distribution and chromosome size range may emerge to reflect changing genomes required for the adaptation of certain phytoplasma groups to varying conditions which include different plant hosts and insect vectors.

Catharanthus roseus plants are the most commonly used host for maintaining phytoplasmas in laboratories worldwide because they are easy to grow, can be readily grafted and display a range of symptoms. However, extraction of high-molecular-mass phytoplasma DNA from infected Catharanthus roseus plants generally gives poor yields such that alternative plant hosts or insects have been used as a richer source of DNA for chromosome extraction.

No restriction endonuclease was found to cut the SPLL-V4 chromosome only once to give a linear molecule. A total of 16 restriction sites were localized on the map, with two of the three EagI sites falling within the two I-CeuI sites. The presence of EagI sites within I-CeuI sites has also been reported for *Borrelia* species (Ojaimi et al., 1994) with the I-CeuI recognition sequence tolerating some degeneracy. There appeared to be no conservation of restriction sites between the SPLL-V4 and WX chromosomes, which have different digestion patterns for all of the four enzymes used. Of the 16 mapped cleavage sites for SPLL-V4, most seem to be fairly evenly spread throughout the genome and this is also the case for the WX map, which contains 20 cleavage sites (Firrao et al., 1996).

Faint bands were often observed in PFGs of SmaI-, EagI-and I-CeuI-digested SPLL-V4 DNA. Based on the estimated sizes of these faint bands, they could not be designated as products from partial digestions. The faint bands were never observed in undigested or BssHII-digested SPLL-V4 DNA, or in preparations of healthy periwinkle. Southern hybridizations using gel-purified faint bands as probes showed hybridization to distinct SPLL-V4 bands, suggesting that these faint bands are of phytoplasma origin, and similarly, SPLL-V4 random probes hybridized with these bands, supporting the conclusion that they represent phytoplasma DNA rather than a different pathogen present in the preparations. If two phytoplasma chromosomes (represented by the strong bands and the faint bands) were present, they would be polymorphic for SmaI, EagI and I-CeuI but not for BssHII. It is unlikely that two separate chromosomes from two phytoplasma strains would have an identical BssHII pattern and different patterns for the other enzymes, so the results indicate the presence of two chromosome types in a phytoplasma population rather than two chromosomes from two different types of phytoplasma. This could occur if there was a chromosome rearrangement, which is a relatively common event in mollicutes (Bhugra et al., 1995; Lysnyansky et al., 1996), within one of the BssHII fragments. This would not alter the BssHII pattern, but would dramatically change the sizes of the fragments generated by the other enzymes. Chromosomal rearrangements have been reported for *Spiroplasma citri* where continued grafting of infected plants resulted in genome polymorphisms compared to *S. citri* passaged by insect transmission (Ye et al., 1996).

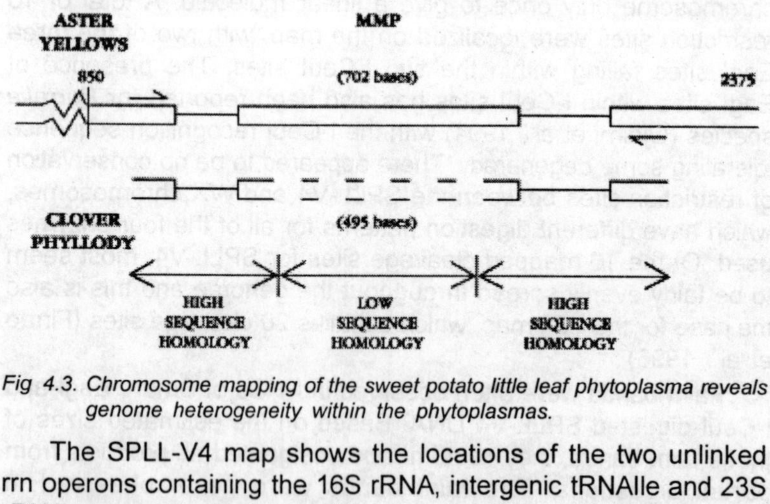

Fig 4.3. Chromosome mapping of the sweet potato little leaf phytoplasma reveals genome heterogeneity within the phytoplasmas.

The SPLL-V4 map shows the locations of the two unlinked rrn operons containing the 16S rRNA, intergenic tRNAIle and 23S rRNA genes, in similar positions to the WX map (Firrao et al., 1996). I-CeuI is a rare-cutting restriction enzyme which recognizes sequences in the 23S rRNA genes of many bacteria (Liu and Sanderson, 1995; Katayama et al., 1995; Toda and Itaya, 1995). The hybridization of only one band in I-CeuI-digested phytoplasma DNA to the 16S rDNA probe implies that both 16S rRNA genes are located within that fragment and therefore that one of the operons is inverted. The inversion of one of the SPLL-V4 phytoplasma rrn loci described in this study has not been reported before for phytoplasmas. The proposed orientation of the phytoplasma ribosomal genes would be consistent with a putative origin of replication (oriC) located between these two genes, somewhere between map positions 433 and 104 in a clockwise direction, with bidirectional replication and transcription. This is the case for *Escherichia coli* where most of the rrn loci are located in the half of the genetic map centred around oriC and is oriented such that transcription occurs in the same direction as replication (Lindahl and Zengel, 1986). Sequence analysis of the entire *M. genitalium* chromosome showed that genes upstream from the origin are transcribed in the antisense direction (Fraser et al., 1995). This is most likely the case for the phytoplasmas.

The only other phytoplasma genes positioned on the SPLL-V4 chromosome map were fus and tuf, encoding the elongation factors G and Tu, respectively. These two genes are linked, as shown by sequence analysis of the PCR product used to generate the probe and as recently shown for other phytoplasmas (Berg and Seemuller, 1999). This is not the case for M. genitalium where the fus and tuf loci are approximately 217 kb apart (Fraser et al., 1995). In other bacteria, including *E. coli* and *Bacillus subtilis*, fus and tuf genes are adjacent (Post and Nomura, 1980; Yasumoto et al., 1996). The tuf locus has been mapped for other mollicutes and usually occurs in association with other important genes involved with DNA replication, repair and transcription (Ladefoged and Christiansen, 1992; Peterson et al., 1995). These genes are usually clustered around the origin of replication. For the SPLL-V4 genome, the clustering of two random probes, the gidA gene, the fus/tuf genes and one of the rrn loci suggests that the putative origin of replication for this phytoplasma, oriC, is upstream of the location of pH80. Once more restriction sites are mapped and more phytoplasma genes cloned and mapped, particularly those associated with DNA replication and repair, transcription and translation, then a better understanding of the location of the phytoplasma origin of replication will emerge.

The only mollicute gene probe, which hybridized to phytoplasma DNA, was pMC5, containing the highly conserved rrn locus, with no other gene probes having sufficient similarity to give a signal. This shows that the similarities at the gene sequence level between mycoplasmas and phytoplasmas are very limited. Even random chromosomal probes from other phytoplasmas did not share sufficient similarity with SPLL-V4 to give a hybridization signal. Further gene mapping of phytoplasmas should therefore focus on genes isolated from the phytoplasma being studied.

These results and the observed lack of restriction endonuclease cleavage site conservation between WX and SPLL-V4 indicated that phytoplasmas are a diverse group of organisms. Even between the two phylogenetically closely related strains SPLL-V4 and TBB there appears to be no conservation of restriction sites, with no common fragments observed between them for any of the four restriction enzymes used. Faint bands were generated for SmaI, EagI and I-CeuI digests of TBB DNA, which were different to those generated with SPLL-V4 DNA. The presence of three bands in I-CeuI digests of TBB DNA may suggest the presence of another chromosome or non-specific digestion, but this needs to be investigated further. The varied restriction patterns between SPLL-V4 and TBB were unexpected because of their close phylogenetic association based on 16S rDNA sequence comparisons, and also because dot-blot hybridizations of TBB and SPLL-V4 DNA with TBB- and SPLL-V4-derived clones indicated that they share a great deal of sequence homology (Schneider et al., 1998). Other comparisons of chromosome restriction patterns between closely related phytoplasmas in the WX phytoplasma group have also revealed polymorphisms (Firrao et al., 1996a). Gene order in several mollicutes is conserved (Peterson et al., 1995; Ye et al., 1995) and this may be a better way to look at inter- and intra-species diversity. To determine whether the same also holds for phytoplasmas, more organisms, closely and distantly related, need to be studied, and more genes need to be mapped. Genetic rearrangements, such as deletions or insertions in particular areas of the phytoplasma genome, may be found to be correlated with biological differences, such as the range of symptoms expressed, host range and transmissibility. Areas of genomic rearrangements found in phytoplasmas could then be targeted for more detailed sequencing studies to identify key genes.

Transmission

Phytoplasmas are completely dependent on their host for existence (obligate parasites). They live in the phloem of plants. So phytoplasmas generally "move" to a new area by grafting, planting infected cuttings, by dodder (a parasitic plant) and by leafhoppers, which migrate annually. In their insect vectors, phytoplasmas reproduce in the hemolymph before traveling to the salivary glands. They are then injected into the phloem of a host while the insect feeds. Transmission from infected to non-infected hosts may also be possible using dodder (*Cuscuta* spp.) to act as a bridge between plants (dodder is a parasitic plant) or via direct grafting of host tissues (Chiykowski 1988). Since Koch's postulates (rules that must be fulfilled to prove that an organism actually causes a disease) are difficult to complete in a non-culturable organism such as phytoplasma, transmission electron microscopy will be utilized to confirm PCR data relative to the success and mechanisms of transmission. Koch's postulates has been completed with a phytoplasma in poinsettia, which causes the sought-after stunting and branch proliferation. There have been many reports of being able to culture phytoplasmas, but none have been repeatable. The major vector involved in transmission is unclear, however the aster leafhopper (*Macrosteleș fascifrons*) would be a prime candidate. Although the potato leafhopper is not strictly a phloem feeder, its sheer numbers and shotgun approach to feeding in alfalfa also makes it a candidate for transmission. Phytoplasmas are phloem-limited, and there is no direct vascular connection to the developing embryo. It has been postulated that, when phytoplasmas change the hormone balance in the plant, they could create that direct vascular connection, but it is unlikely the phytoplasmas could survive the desiccation usually associated with seed formation. Like mycoplasmas, phytoplasmas are organ/tissue specific to an extent. They like roots very well, but can be found in many places in the plant.

Circulifer tenellus, inoculative for the beet leafhopper transmitted virescence agent (BLTVA), were fed on young *Arabidopsis thaliana* ecotype La-O. After a period of five weeks, the plants developed disease symptoms including virescence, phyllody and the proliferation of flowers. Transmission electron microscopy revealed the presence of mycoplasma-like bodies in the phloem of symptomatic plants. *A. thaliana* has been demonstrated to be a host of BLTVA, a plant pathogenic mycoplasma-like organism.

Fig 4.4. Dodder transmission of phytoplasma from an infected periwinkle plant to a virus-free restricted-branching poinsettia plant.

In addition to their importance in causing damage to crop plants, phytoplasmas are of interest because of their profound effects on morphology and physiology of their hosts. One group of phytoplasmas, referred to as the virescence agents, cause a wide range of symptoms, which include phyllody (leaf-like flower petals), virescence (greening of normally pigmented floral organs), proliferation (the development of shoots or flowers from adventitious meristems), stunting, stem elongation and seed sterility. Many of these host symptoms are similar to the responses that might be elicited by the exogenous application of plant growth substances.

One phytoplasma the beet leafhopper transmitted virescence agent (BLTVA), has been demonstrated to induce flowering in plants grown under environmental conditions which do not permit non-inoculated control plants to flower (Golino, 1988). The nature of this host induction response (HIR) remains to be determined. Plant growth substances are very likely involved. It is conceivable that this phytoplasma contains genes, which affect the expression of host plant hormone metabolism, or, possibly, is a source of prokaryotic genes for synthesis of GAs. It is unfortunate that this organism has not as yet been cultured axenically, since this would allow use of the techniques of molecular biology to determine the mechanism of the HIR. Research is needed to determine the vector species involved in transmission and the timing of transmission events.

Pathogenesis

Plant-pathogenic mycoplasmas are widespread disease agents, of which the phytoplasmas are the most important group. As phytoplasmas cannot be cultured under axenic conditions, they are difficult to study. Thus, it is not known, how they induce disease. For this reason, a culturable plant-pathogenic acholeplasma (*A. laidlawii* var. *granulum*), which is closely related to the phytoplasmas, has been selected as a model to investigate phytoplasma pathogenicity. It could be shown that there exist two different forms of RNA polymerases (RNAPs) in this pathogen, of which one is not present in saprophytic types of *A. laidlawii* or in other mycoplasmas not pathogenic to plants. It is supposed that the additional RNAP transcribes genes whose products are relevant for disease induction. There is evidence that a fructose 1,6-diphosphate-specific lectin is such a product that plays a major role in pathogenicity. This lectin may mediate attachment of mycoplasmas to host membranes and seems to deregulate photosynthesis by binding of fructose 1,6-diphosphate. The depletion of this compound results in an uncontrolled synthesis of starch, which, in turn, leads to an enormous accumulation of this carbohydrate and the impairment of chloroplast function.

The discovery of mycoplasmalike organisms (MLOs) as the probable cause of plant "yellows diseases" in 1967 was followed by a steady growth in knowledge of these unique microbes and their relatives (members of the class Mollicutes). This growth has been punctuated by the discovery that cell wall-less prokaryotes infecting plants were of two distinct types, mycoplasmalike organisms (MLOs) and spiroplasmas; by the isolation of spiroplasmas in culture, proof of their plant pathogenicity, and description of over 30 Spiroplasma species, many from insects; by discoveries of MLOs as the probable causes of "yellows" diseases worldwide in more than 300 plant species; by recognition of the evolutionary position of MLOs within the class Mollicutes, determination that MLOs comprise a monophyletic clade in the class, and acceptance of the term phytoplasma for members of the clade; by the development of rapid and sensitive means for phytoplasma detection and identification; by the construction of schemes for classification of phytoplasmas based on molecular data; by the emergence of a provisional phytoplasma taxonomy and the designation of "Candidatus Phytoplasma species"; by new knowledge gained about the ecology and molecular biology of spiroplasmas.

These and other advances have yielded fundamental insights into the nature of the phytopathogenic mollicutes and raised hopes that knowledge to be gained of these microbial pathogens could lead to effective and environmentally safe ways to control diseases that they cause. At the same time, it has become increasingly apparent that much remains to be learned.

Studies to unveil the mechanisms of pathogenicity in phytopathogenic mollicutes are currently underway. Studies such as these, probing plant responses to infection through investigation of plant gene expression and mutagenesis of the phytopathogens, will provide valuable insights. Other approaches could also prove useful to understand pathogenicity of phytopathogenic mollicutes. One such approach is genomics; phytoplasmas are noted below as one example, but, of course, genomics offers a powerful tool for better understanding of spiroplasma pathogenicity as well.

Just as phytoplasma genomics could reveal both the nature of sequences presumably lost, or modified, in the evolution of phytoplasmas from ancestors comparable to known acholeplasmas, it could also reveal the nature and origin of sequences that may have been acquired by phytoplasmas from other organisms. For example, it is possible that genes responsible for their plant pathogenicity were acquired by phytoplasmas through horizontal gene transfer. Distantly related phytoplasmas can cause apparently identical symptoms in certain plant species, possibly because phytoplsmas as a group possess relatively few mechanisms for plant pathogenicity. This line of reasoning led Davis and Sinclair to hypothesize "...a common ancestral origin of phytoplasma pathogenicity genes...". Genomics would provide one avenue for testing this hypothesis.

Future studies in phytoplasma and spiroplasma genomics offer avenues for aiding the quest to understand specificities in insect transmission and mechanisms accounting for the emergence of new plant diseases and resurgence of "old" diseases; for developing improved methods for phytoplasma detection and identification; and for identifying molecular targets for manipulating and controlling host-pathogen interactions.

Diseases

Phytoplasmoses can cause significant economic damage. The chronic presence of phytoplasma in alfalfa would also have implications for other crops. Aster yellows phytoplasmas have significant agronomic effects on a wide variety of crops including carrot and lettuce and barley and can cause significant losses in crops such as cucurbits, celery, onion, and legumes. In addition, phytoplasmas have been associated with slow decline of various shrubs and tree species. They are epidemic and have a long incubation period. Symptoms on infected trees become evident only one or two years after contamination, when the disease has already spread to other trees or orchards. Disease symptoms vary between the species, but usually the leaves show unusual yellowing or reddening and fall off prematurely leaving bare branches. The branches die back from the tip, leading to eventually to plant death. Once initial symptoms appear, plants seldom recover. Disease distribution matches that of the passionvine planthopper, one of the prime suspects for spreading the disease. Phytoplasmas (previously called MLOs) are important plant pathogens as they are associated with diseases in several hundred-plant species worldwide. Diseases associated with the presence of phytoplasmas in phloem, typically exhibit an array of symptoms suggestive of profound disturbances in the normal balance of plant hormones. Symptoms in plants include sterility of flowers, virescence (the development of green flowers and loss of normal flower color), phyllody (the development of floral parts into leafy structures), proliferation of axillary shoots or auxiliary

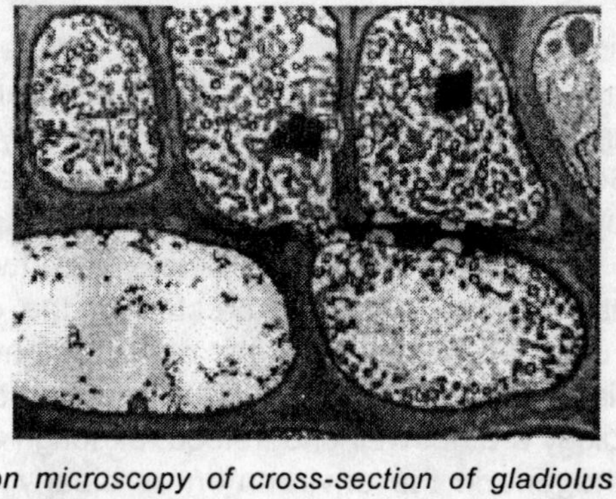

Fig 4.5. Electron microscopy of cross-section of gladiolus phloem tissue infected with phytoplasma.

shoots resulting in witches'-broom appearance, generalized stunting and decline, bunchy appearance of growth at the stems, yellowing, phloem necrosis, and dieback of branches in woody plants. In general, the range of symptoms attributed to phytoplasmal infection are indicative of detrimental effects on plants. The damage due to the growth disorders induced by phytoplasmal infections ranges from partial reduction in yield and quality to nearly total loss of the crop.

Alfalfa plants store carbohydrates (starches and sugars) in their roots and crowns. These carbohydrate reserves are used for regrowth in the spring and after each cutting. Carbohydrate reserves are replenished when the plant reaches 6 to 8 inches tall. Management practices such as cutting too frequently can deplete carbohydrate reserves and lower stand persistence. Similarly, the presence of phytoplasma in the phloem of infected plants could affect the movement and available amount of carbohydrate reserves for the plant leading to reduced productivity and stand persistence. The presence of phytoplasma over several years in a field could be partially responsible for the observed decline that often occurs in alfalfa plantings over time. It is often noted that such decline is not associated with any of the traditional alfalfa pathogens and in the past has been blamed on variety genetics or environmental conditions. It is a hypothesis that phloem-limited phytoplasma are having a delayed yet significant impact on forage yield due to utilization and interruption with transport of carbohydrates. Such an impact would only be noticed over time.

Host Range

Exploration of the interactions of phytopathogenic mollicutes with their plant and insect hosts is a fascinating and very active field of research as we move toward the 21st century. The attention is justified. Severe diseases of food, fiber, and ornamental plants are caused by insect-transmitted spiroplasmas and phytoplasmas, resulting in significant economic losses worldwide. Disease management practices, whether directed at the control of the pathogen or its insect vector, or at the development of resistance in the host plants, have been limited. In addition, mollicute biology and epidemiology are not well understood, although the distribution of these pathogens in the environment is clearly dependent on the movement and host selection of their insect vectors. Little is known about the specific events at the microbe-vector interface that result in mollicute traversal of the insect and define the nature of transmission specificity. Information in each of these areas is basic to the improvement of our approaches to management of the plant maladies caused by this group of unusual prokaryotes.

A new stolbur phytoplasma vector, *Pentastiridius beieri*, had been found in sugar beet crops affected with SBR (Syndrome des basses richesses). The vectoring ability and biology of this new vector were thoroughly studied. It is particularly interesting that the species belongs to the Cixiidae family, together with Hyalesthes obsoletus, formerly known as a stolbur and Grapevine Bois noir diseases, and *Oliarius atkinsoni*, a vector of PYL (a phytoplasma close to stolbur) in New Zealand.

However, it was shown that SBR may be caused either by the stolbur phytoplasma or by a BLO (phloem-restricted Bacterium-like organism). Both these phloem bacteria were transmitted by the same insect species. Both phytoplasma and BLO could be detected in the same sugar beet taproot.

Assays were set up to verify if *Metcalfa pruinosa*, a polyphagous leafhopper (Flatidae) now colonizing Italy and the south of France, might be a phytoplasma vector. The results suggest that acquisition of phytoplasmas was effective, but that they did not multiply and move in the body of the insects, since PCR signals decreased after the transfer of phytoplasma to healthy plants and no transmission was obtained. However, insects did not survive after three weeks of experiments and additional trials should be done.

A procedure for the demonstration that insects can be phytoplasma vectors was developed and used in laboratory conditions with reared phytoplasma-infected leafhopper vectors and with wild insects captured in the field. Instead of using test plants for the demonstration of transmission, insects are fed through a parafilm membrane into a TE-sucrose diet. DNA was extracted from the diet after feeding, and PCR-tested with ribosomal primers to detect the presence of phytoplasma that would have been injected with saliva into the diet share. The data fitted the known cycle of phytoplasma in the insect body after acquisition, with a 4 weeks latency before ability to transmit. Early detection in the first days after acquisition was assumed to be due to phytoplasma that were adsorbed to the stylet. Control with healthy insect reared in the insectary was negative. The procedure showed to be reliable to demonstrate infectious status of insects from the field according to the period of capture.

Host resistance will be an issue of great interest in the early years of the coming millenium. Model insects such as the fruit fly, *Drosophila* sp., and the tobacco hornworm, Manduca sexta, utilize phagocytic cells and defense-related hemolymph proteins such as defensins and cecropins. Since surface molecules of invading walled bacteria may induce the genes of some insect immunity-related molecules, it will be interesting to test whether they also are activated by wall-less prokaryotes. Despite a growing body of information on immune reactions in a few insect species, little is known about immunity-related cells and proteins in the leafhoppers and planthoppers that transmit phytopathogenic mollicutes like phytoplasma. Transcription by mycoplasmas of different versions of surface protein genes facilitates avoidance of the animal/human immune recognition or other adaptations to the environment of the vertebrate host. Multiple copies of certain genes are found also in the genomes of several spiroplasma strains; the P58 putative adhesin of *S. citri*, for example, has three similar copies in BR3-3X. Whether such repetition at the gene level reflects alternative expressed forms of the protein, or is related to immune avoidance, is not yet known.

Identification

A preliminary identification can be made with staining methods. Two methods that can be used are the DAPI method and the Deines' staining method. DAPI (4,6-diamidino-2-phenylindole) is a nucleic acid-specific stain, and stain bacteria in the phloem (Sinclair 1989). Phytoplasmas that stained by Dienes' stain metabolize and creates a blue colour. Healthy phloem should be clear of colour (Deeley et. al. 1979). A drawback to both of these methods is a lack of specificity. Dienes' stain is more specific than DAPI, but does stain callose deposition. Using microscopy, phytoplasmas can also be detected using scanning electron microscopy and, more commonly, transmission electron microscopy. Other detection methods include DNA probes, monoclonal and polyclonal antibodies, and PCR (Chen et. al. 1993, Lin et. al. 1986, Chen and Jiang 1990, Lee et. al. 1998). There are currently 14 groups of phytoplasma which have been distinguished by 16s rDNA RFLP analysis. RFLP analysis of the tuf gene may provide even finer resolution (Schneider et. al., 1997).

Phytoplasma must be maintained in the host. This can be done in plant tissue culture, continuous graft or insect transmission, or freezing leafhoppers (Bertaccini et. al. 1992, Lee and Chiykowski 1963.). All of these methods have an element of fragility.

Since phytoplasma are so hard to work with, it has been extremely challenging to complete Koch's postulates (proof of pathogenicity) with them. The closest has been Ing-Ming Lee's work with poinsettia's (1997). The greatest factor that has led scientists to believe a phytoplasma is causing a disease is frequent association with diseased tissue. In some cases it has been suspected that other organisms such as bacteria are actually causing the disease and not the phytoplasma, but Koch's postulates have not been completed with these systems yet (Davis et. al. 1996).

Much as with viruses, many plants can be cured of phytoplasmas with heat treatments and/or by passing them through tissue culture (Kunkel 1941). This is a very difficult process, and it's easier to pass the phytoplasma-infected plant through a seed cycle, since phytoplasmas are not seed-transmitted. Phytoplasmas are phloem-limited, and there is no direct vascular connection to the developing embryo. It has been postulated that, when phytoplasmas change the hormone balance in the plant, they could create that direct vascular connection, but it is unlikely the phytoplasmas could survive the desiccation usually associated with seed formation. Like mycoplasmas, phytoplasmas are organ/tissue specific to an extent. They like roots very well, but can be found in many places in the plant.

The use of the polymerase chain reaction (PCR) for amplification of ribosomal RNA and ribosomal protein gene sequences, and the use of restriction fragment length polymorphism (RFLP) and/or sequence analysis of amplified DNAs are now widely used, standard methods for phytoplasma identification. Phytoplasma pathogenicity genes will eventually supplant the use of ribosomal RNA, ribosomal protein, and uncharacterized sequences in the detection and identification of phytoplasmas. Computerized analyses and comparisons of RFLP patterns have been made feasible through computer programs developed during the past several years. So it can put the hope that microchip DNA arrays and "real time" PCR become widely used for the detection and identification of phytopathogenic and other mollicutes.

Genetic Transformation of Phytoplasmas and Expression of Phytoplasma Genes

Genetic transformation of phytoplasmas and investigations of phytoplasma gene expression have not yet been realized. Nevertheless, these research avenues would provide useful complements for future studies of the phytoplasma genome and aid understanding of phytoplasmas. Reports of the expression of phytoplasma genes in walled bacteria should be forthcoming within the next year or two; expression of selected phytoplasma genes in plants will likely follow. Conceivably, expression of phytoplasma genes could also be achieved in other, cultivable members of the class Mollicutes to aid in analyses of gene function; expression of some foreign genes has been investigated in non-phyto-pathogenic mollicutes. Although phytoplasmas are not presently cultivable in cell-free media, it is conceivable that genetic transformation of phytoplasmas can, and eventually will, be achieved even in the absence of isolation of these mollicutes in pure culture. Such experiments would be highly valuable in identifying phytolasma genes that determine specific phenotypes. Several candidate experiments for future research have been outlined. Such experiments could lead to an understanding that will permit beneficial manipulation of host-mollicute pathogen interactions.

Phytoplasmas in Cellular Environments

Some of the most interesting aspects of the biology of phytopathogenic mollicutes involve the limitation of phytoplasmas and spiroplasmas to phloem tissue in infected plants and their transmission only by leafhoppers and related phloem-feeding insects and, rarely, plant bugs. What limits these pathogens to phloem and, apparently, specifically to sieve cells? Thus far, there has been no report of phytoplasma or spiroplasma transmission by a phloem-feeding aphid; the reasons for lack of transmission by aphids are not known. Sites of mollicute attachment to insect tissues and other pathogen-insect interactions can be cited in a general sense to explain transmission specificities. But till it is questionable what molecular mechanisms, that are present in leafhoppers and presumably absent in aphids, account for the differences in mollicute transmission between these major insect groups?

Evolution

The study of phylogenetic relationships of spiroplasmas and phytoplasmas has been aided by the statistical construction of phylogenetic trees based on the analysis of sequences of conserved genes such as ribosomal RNA genes and ribosomal protein genes. These trees provide a convenient framework for hypotheses concerning the relationships among distinct lineages of these mollicutes. They also provoke questions concerning, for example, phytoplasma evolution and the emergence of new biotypes and of new, distinct species. Possible biological events in the development of distinct phytoplasma lineages, including the roles of monophagous and polyphagous insect vectors and of expanded plant and insect host ranges, have been discused by Lee et al. Current knowledge still leaves interesting questions unanswered. For example, by what molecular processes do these events occur, and to what extent are they occuring now?

Both non-phytopathogenic and phytopathogenic mollicutes are found in associations with plants; in their associations with plants, non-phytopathogens are known only from plant surfaces, whereas, phytopathogens are known only from plant interiors. The association of spiroplasmas and nonhelical mycoplasmas with the surfaces of plants, including flower parts, provokes other questions concerning contemporary, as well as past, evolutionary connections between plant-inhabiting phytoplasmas and those of arthropods and vertebrate animals. Did, or do, interactions of plants with ancestral or present day non-phytopathogenic mollicutes afford conditions favorable for mollicute introductions into new ecological niches and and favor the divergent evolution of distinct Spiroplasma, phytoplasma, and Mycoplasma lineages?

That insects and other arthropods regularly visit flowers and feed on or in these and other plant organs is well known, but vertebrate animals also visit flowers, where they eat flower parts or feed on nectar. "Conceivably, flowers could serve as foci for dissemination of pathogens among both arthropods and vertebrates, thus providing an important common ground for transfer of widely different agents between divergent groups of fauna". Similarly, vertebrate animals consume other plant parts, on the surfaces of which may occur mollicute species deposited by insects, and some insects that simply bite or actually feed upon living vertebrates also feed on plant nectar. Could plant-arthropod-vertebrate associations have played roles in the evolution of mollicutes, including vertebrate pathogens? Or, perhaps more correctly stated, what roles did plant-arthropod-vertebrate associations play in the evolution of species in the class Mollicutes? And, are any present day, vertebrate-inhabiting mollicutes disseminated through the vehicle of plant parts and/or plant-visiting arthropods?

Ecology

It is well known that phytoplasmas infect plant species in natural ecosystems and that such plants can serve as sources of phytoplasmas for infections of plants in agro-ecosystems. The effects of phytoplasmas on agro-ecosystems have been and are subjects of numerous studies. It would be of interest to know more about phytoplasmal infections of plants in natural ecosystems and about the overall consequences of these infections on the natural environment.

In natural ecosystems, some plant species exhibit obvious disease symptoms in response to phytoplasmal infections. In such cases, environmental consequences of phytoplasmal infections may be clear, as for example, in cases of decline of ash trees due to ash yellows disease and of death of American elm trees due to elm yellows disease. However, it is apparent that some species in natural ecosystems exhibit no obvious symptoms of phytoplasma infection under normal conditions. For example, symptomless plants of wild grapevine (*Vitis riparia*) harbor phytoplasma strains that are associated with development of severe disease symptoms in cultivated grapevine (*V. vinifera*). Trees of black locust (*Robinia pseudoacacia*) infected by phytoplasma in eastern USA generally exhibit no disease symptoms unless the tree is cut to leave a stump; then, witches' broom growths sprout from stump and roots, symptoms characteristic of the black locust witches' broom disease. In apparent phytoplasmal infections may be common in nature and may have subtle effects on plants in natural ecosystems, perhaps rendering them more susceptible to damage by other stresses.

This First Internet Conference on Phytopathogenic Mollicutes marks three decades of remarkable progress toward understanding plant "yellows diseases" and the causal pathogens of these numerous and worldwide maladies. This progress presents new challenges, the pursuit of which is encouraged and facilitated by continuing developments in this and other fields of biology.

Taxonomy

For many years, viruses were accepted as the likely cause of numerous yellows-type diseases of plants, a conclusion based upon symptomatology, mode of transmission by insects, and patterns of multiplication in plants and vector insects. Typical of viruses, the yellows agents were filterable, many were graft-transmissible, and some were shown to interfere with each other in plants and insects. Thus the proposal in that the etiological agents of these diseases could be wall-free prokaryotes rather than viruses caught the field of plant pathology totally by surprise. This landmark discovery of a novel and fundamentally different group of plant pathogens prompted a comprehensive reassessment of other yellows diseases and led to similar findings of pleomorphic, wall-less prokaryotes in the phloem of many other affected plant species.

The term mycoplasma-like organisms (MLOs) was first used to denote these microoganisms due to their morphological and ultrastructural resemblance to mycoplasmas. Like the latter cultivable Mollicutes, MLOs were shown to be sensitive to tetracycline but not to penicillin antibiotics. However, in contrast to mycoplasmas, which cause an array of disorders in animals and humans, the phytopathogenic MLOs resisted all attempts to culture them in vitro in cell-free media. Profound biological differences between MLOs and mycoplasmas were also indicated by the fact that MLOs were transmitted to and spread among plants in nature by phloem-feeding vector insects, mostly leafhoppers, planthoppers and psyllids.

In the three decades since their discovery, and based on evidence provided by fluorescence and electron microscopy or transmission experiments, MLOs have been implicated as the etiological agents of several hundred plant diseases worldwide, including numerous diseases of economically important food, fiber, forage, fruit and timber. Only within the last decade, following the application of molecular technologies on these diseases, was the enigmatic status of MLOs amongst the prokaryotes clearly resolved. Evidence derived from these studies validated the long-standing assumption that MLOs are indeed Mollicutes and showed them to be only distantly related to true Mycoplasma species. This revelation led to a proposal that the term mycoplasma-like organism be replaced with the generic name of phytoplasma, a designation that clearly reflects their plant host association. Furthermore, molecular data derived from assays employing serological and DNA probes or DNA amplification by the polymerase chain (PCR) reaction have since provided considerable insights into the diversity and genetic interrelationships of phytoplasmas. Sequence or restriction fragment length polymorphism (RFLP) analysis of ribosomal RNA (rRNA) gene sequences amplified by PCR provided the basis for several comprehensive studies on phytoplasma phylogeny and taxonomy.

Phytoplasma differentiation and classification have been an underlying or primary theme in several recent reviews authored by research groups that have led the quest in search of a valid taxonomy for this group of plant pathogens. These reviews were used as the primary sources for the ensuing introductory statement, the intent of which is to provide a synopsis of the recent history and current status of phytoplasma taxonomy.

Several "Candidatus Phytoplasma species" have been named, and the naming of others that have already been recognized as new lineages is expected during the next year or so. However, the known phytoplasmas are likely to represent a small fraction of the genetic diversity and distinct lineages among existing phytoplasmas worldwide. During the next thirty years, it is feasible that described and named lineages will number far more than the less-than-twenty that are currently recognized as worthy of recognition as separate "Candidatus Phytoplasma species". The reasons for this hypothesis are several. First, vast regions of the world have not been studied to detect and characterize phytoplasmas present in agricultural crop plants. Further, it is likely that many plant species, that form part of the natural vegetation, harbor phytoplasmas without obvious evidence of disease; few such plants have been examined for the presence of phytoplasmas. Also, analysis of highly conserved genes, such as 16S rRNA genes, and studies of speciation in other bacteria, indicate that analysis of some conserved genes may not distinguish between closely related species. In the light of this last observation, it may be asked, which more variable genes will be found suitable for making finer distinctions among closely related species? .

As new and expanded information will be gathered from studies of the phytoplasma genome, it is likely that genomics will play an increasingly important role in the future of phytoplasma taxonomy. The same prediction could be made for the taxonomy of other mollicutes. For example, when fully understood, the completely sequenced genome of a phytoplasma, spiroplasma, or other mollicute, will define its genotype and reveal its phenotype. In current, formal taxonomy of Mollicutes, phenotypic and some genotypic description is required for descriptions of new species. Complete inference of phenotype from genotype that is from full genome sequence data is undoubtedly a goal for the distant future. However, eventually, full understanding of a completely sequenced genome would yield more precise and far more detailed species descriptions than afforded today. The rapid advance in sequencing capabilities and in computer programs for analyses of large sequences suggests that, in the future, sequencing of the entire genome of a prokaryote of 2mb genome size or less will become relatively inexpensive and routine. It is reasonable to speculate that the molecular based taxonomy of phytoplasmas--currently based heavily on 16S rRNA gene sequences--will be improved and that, eventually, genome studies and other research approaches will reveal families of genes that will be highly useful for distinguishing phytoplasmas at genus, species, and lower taxonomic levels. Such genes (and their putative or assayed products) would present opportunities for expanded descriptions of distinct phytoplasma lineages, as well as for improved means for rapid and accurate detection and identification of phytoplasmas.

Based on dissimilarities of restriction sites detected by rDNA RFLPs, a total of nine primary 16S rDNA groups (termed 16Sr groups) and 14 subgroups were recognized in a classification scheme. Phytoplasma groups delineated by these analyses were consistent with genomic strain clusters previously identified by DNA probe hybridization although a greater diversity among strains comprising group 16SrI (aster yellows and related strains) was indicated from the hybridization data. More detailed subdivision of 16SrI and 16SrIII (X-disease and related phytoplasmas) groups was revealed after combining RFLP data derived from both the 16S rRNA gene and more variable ribosomal protein genes. This strategy for finer sub-group differentiation was used to modify and expand upon earlier classifications to incoporate many new phytoplasma strains identified within the last few years. At least 14 primary 16Sr groups consisting of 16SrI, aster yellows; 16SrII, peanut witches' broom; 16SrIII, X-disease; 16SrIV, coconut lethal yellows; 16SrV, elm yellows; 16SrVI, clover proliferation; 16SrVII, ash yellows; 16SrVIII, loofah witches' broom; 16SrIX, pigeonpea witches' broom; 16SrX, apple proliferation; 16SrXI, rice yellow dwarf; 16SrXII, stolbur; 16SrXIII, Mexican periwinkle virescence; 16SrIV, Bermuda grass white leaf, and at least 32 strain subgroups have been recognized. A total of 45 subgroups were identified when ribosomal protein gene RFLP data were also considered in the analyses.

DNA amplification by PCR, thermal cycling protocols for direct sequencing of amplification products, and gel additives permitting extended sequence reads have provided ready access to phytoplasma 16S rRNA gene sequences at increasingly modest cost. The first major attempts to sequence amplified 16S rRNA genes either directly, or after cloning into suitable vectors, were undertaken. From these collective efforts, at least 30 near full-length sequences were compiled from a diversity of strains previously characterized by rDNA RFLP analysis, and resulted in the first comprehensive phytoplasma phylogeny As was deduced from earlier studies, these more recent analyses provided further compelling evidence that phytoplasmas constituted a unique, monophyletic clade of Mollicutes more closely related to Acholeplasma than to the true Mycoplasma species. Within the phytoplasma clade, as many as 13 major subclades (groups) represented by the following type strains (1) stolbur; (2) aster yellows, (3) apple proliferation, (4) coconut lethal yellowing, (5) pigeonpea witches' broom, (6) X-disease, (7) rice yellow dwarf, (8) elm yellows, (9) ash yellows, (10) sunnhemp phyllody, (11) loofah witches' broom, (12) clover proliferation and (13) peanut witches' broom were recognized.

Twenty distinct phytoplasma groups, including eight novel groups absent from previous classifications Australian grapevine yellows (AUSGY), Italian bineweed stolbur (IBS), buckthorn witches' broom (BWB), Spartium witches' broom (SpaWB), Italian alfalfa witches' broom (IAWB), Cirsium phyllody (CirP), Bermuda grass white leaf (BGWL) and Tanzanian lethal decline (TLD) were recently defined. Following an analysis of 57 full-length or near full-length 16S rDNA sequences, these new phylogenetic groups were delineated on the basis of tree branching patterns and by differences in sequence similarity of at least 2.3%. Within the framework of this latest analysis, 189 additional phytoplasma isolates for which other DNA-based or serological data were available were assigned to each of the 20 primary phylogenetic groups. In light of these additions, groups such as aster yellows, faba bean phyllody, X-disease, sugarcane white leaf and apple proliferation were quite heterogenous, indicating that speciation based solely upon 16S rRNA may lead to an oversimplification of phytoplasma diversity.

The spacer region (SR) separating the 16S from the 23S rRNA gene of phytoplasmas was also shown to be a reliable phylogenetic marker. Phylogenetic trees derived from the entire 16-23S SR or variable regions flanking the tRNAile gene differentiated phytoplasmas into groups concordant with the major groups previously established from analyses of 16S rRNA genes. Several benefits to using 16-23S SR sequences for the purpose of phytoplasma classification were also realized. Since the total length of the sequence is about 250 bp, a full-length sequence through this region can be readily obtained in both directions with only two sequencing primers. Direct sequencing from PCR products is often easier because of higher yields with smaller PCR products and the fact that sequences are inherently more variable than 16S rRNA genes. Phytoplasmas differ in their 16S rDNA sequence by no more than 14%, whereas their respective 16-23S SR differed by as much as 22%. This added variation may contribute to improved accuracy of phytoplasma classification at the subgroup level.

In response to a rapidly growing database of phylogenetic markers, the Working Team on Phytoplasmas of the International Research Programme of Comparative Mycoplasmology (IRPCM) proposed that taxonomy of phytoplasmas be based upon 16S rRNA genes. In this approach, major groups delineated by phylogenetic analysis of near full-length 16S rRNA gene sequences are considered to represent distinct species. This proposal was agreed to and adopted as policy by the ICSB Subcommittee on the Taxonomy of the Mollicutes, which also recommended that the "Candidatus" status proposed in be used for assigning names to Phytoplasma species. Thus far, four Candidatus species have been described: *Phytoplasma aurantifolia*, *P. australiense*, *P. australasia* and *P. fraxini*.

5

SPIROPLASMA

For decades following the original descriptions of the disease (Kunkel, 1946,1948), corn stunt, as well as other so-called plant "yellows" diseases, were presumed to be caused by a virus. In 1968, pleomorphic cell wall-less microbes (mycoplasmalike organisms, MLOs) were reported in diseased corn (*Zea mays* L.) plants and in insect vectors capable of transmitting the disease (Granados, 1969; Maramorosch et al. 1968). Spiroplasmas were discovered in 1972 when helical microorganisms were envisioned by phase microscopy from sap of plants infected with corn stunt disease. In 1973, Davis and Worley, in the absence of cultivation of the infectious agent, coined the trivial term "spiroplasma" for the helical motile filamentous mycoplasma-like organisms they had observed in association with corn stunt disease. A similar organism had already been cultivated in separate laboratories in California and France. Other spiroplasmas, discovered earlier and studied under various aliases, then turned up. One was from *Drosophila* (sex-ratio organism) and two were from ticks (strain 277F and the suckling mouse cataract agent). Of the unnamed spiroplasmas, only one (strain 277F) was readily cultivable. In 1975, two research groups achieved cultivation of the corn stunt spiroplasma, but the cultured organism showed partial serological relationship with *S. citri.* At the same time that the citrus stubborn causative agent was being described another plant disease was also being investigated as a member of this new genus of bacteria. In 1975, cultivation of the corn stunt spiroplasma in artificial media in vitro permitted conclusive evidence that it was the cause of corn stunt disease and a description of the second member of the Spiroplasma genus, *Spiroplasma kunkelii* was published. In retrospect, three previously observed microorganisms were subsequently determined to belong to the genus Spiroplasma including the sex ratio organism of *Drosophila*; an isolate from a rabbit tick designated as strain 277F; and another rabbit tick isolate referred to as the suckling mouse cataract agent or strain SMCA.

By 1975, the clear association of Spiroplasmas with arthropod hosts was clearly established, as was their ability to cause pathology in some plant hosts. However, in 1977 a link between Spiroplasmas and arthropod disease was also established by Truman Clark's work on a honeybee disease that had the ability to be lethal. When looking for the reservoir of these organisms several different strains of Spiroplasmas were detected on the surfaces of flowers. The first honeybee spiroplasma, *Spiroplasma melliferum* and the first floral spiroplasma, *Spiroplasma floricola* were described shortly thereafter.

Spiroplasmas are present in the hemolymph of almost all insects; there probably are several million strains. They can also cause diseases in plants but are usually associated with a vector. For example, a leafhopper carries a spiroplasma that infects orange trees Spiroplasmas are similar to mycoplasmas. They do not have a cell wall (cell wall deficient) and have among the smallest genomes of any living organism. Mycoplasma are the smallest and perhaps the oldest life form. These bacteria, one cause of "walking pneumonia", are thought by many to be rather fragile, but nothing could be further from the truth. They tolerate extreme fluctuations in temperature, lay dormant in the soil for generations and survive the harshest elements; only drano-like chemicals kill them effectively outside the body. Under normal circumstances our immune system efficiently deals with this complicated, membrane enclosed piece of DNA. A common phenomenon among the mycoplasmas is that the organisms bind host proteins that often are of identical molecular weight to their surface proteins and, therefore, are looked at by the immune system as being the same as the host. The spiralin protein on the surface of spiroplasmas shows a migration pattern on gel electrophoresis with a molecular weight of 27,000 Da to 30,000 Da, similar to that of the prion protein. This biochemical similarity is compatible with spiroplasma etiology.

Taxonomically, Spiroplasmas are in the Domain Bacteria, Phylum Firmicutes, Class Mollicutes, Order Entomoplasmatales, Family Spiroplasmataceae, and Genus *Spiroplasma*. The Order Entomoplasmatales is split into two families, one containing the *Entomoplasmas* and *Mesoplasmas* and the other containing the *Spiroplasmas*.

General Characteristics

Characterization of Spiroplasmas requires the determination of morphological, biological and serological properties. Preliminary characterization of triply cloned Spiroplasma isolates involves microscopic observations of morphological characteristics. Dark-field microscopic observation allows for observation of motility and helicity. Electron microscopy reveals the lack of a cell wall and dimensions approximating 200 nm in width and a length range of 3-12 µm.

Biological properties

Determination of optimal temperature for growth. All Spiroplasmas have the ability to be filtered through 220 nm pores membranes. They have ability to ferment glucose. They are resistant to 500 U of penicillin/ml. They require cholesterol for growth. They are unable to hydrolyze urea. Variable characteristics include ability to hydrolyze arginine and optimal growth temperature. Despite all of these different biological characteristics exhibited by Spiroplasmas, there is no single genus-specific property that distinguishes the Spiroplasmas from all other mollicutes. Metabolically, some questions are still unanswered about Spiroplasma like nutritional uptake and energy metabolism, nutrient transport, DNA repair mechanisms, and regulation of replication, transcription and translation. Specific work is being continue through various experiments using genomic analysis, which can give the answers how do mollicutes maintain cellular homeostasis; which genes are involved with pathogenesis; how is the genome content organized; and, how will Proteome analysis provide information about gene expression.

Molecular properties

G + C base composition, genome size and 16S rDNA sequence analysis are the basic molecular characteristics of Spiroplasmas. The only requirement for a new species description is the determination of G + C base composition (mol%). G + C base composition ranges from 24-31% for Spiroplasmas and is performed using the melting temperature method. The genome size ranges from 940-2,240 kbp for Spiroplasma sp. and is determined using pulsed-field gel electrophoresis. 16S rDNA sequence analysis is considered valuable for phylogenetic analysis and for the unambiguous taxonomic placement of a novel isolate within the Spiroplasma genus. The 9.6 kbp stretch contains a gene for a 58 kDa membrane protein that has some sequence similarity to adhesions of mycoplasma. This P58 and another protein candidate are being evaluated for involvement in the attachment of spiroplasma to the insect gut and salivary gland. The genome has three P58 gene copies. The other two are being isolated and characterized. The characterization should reveal whether the multiple copies play a role in growth in three different environments like free, in plants and in insects.

Serological properties

Currently the most important criterion in the characterization of new spiroplasma isolates is through the use of techniques to determine serological relatedness. The deformation test (DF) is universally used in Spiroplasma characterization. The DF test screens new isolates against antisera to representatives of existing group strains. Isolates that exhibit a DF titer of greater than 1:40 with established spiroplasma groups are considered candidates for serovar (or subgroup) status. Reciprocal DF tests are then usually performed to clarify the serological relatedness between the two (or more) cross-reacting strains. Homologous reactions usually have endpoints at 1,280 to 20,240, however, some may be as low as 320-640 and still indicated relatedness. Partial relatedness between two strains shows titers that are lower than the homologous titers. The metabolism inhibition (MI) test is usually the second serologic technique used. The MI test is more sensitive than the DF test and is important in showing subgroup or serovar relationships. The growth inhibition and enzyme-linked immuno-sorbent assays are also occasionally used as a secondary serological assay after the DF test has been performed.

Corn Stunt Spiroplasma

It is a helical shaped, motile, organism. This prokaryote lacks a cell wall and belongs to the genus Spiroplasma. Corn Stunt Spiroplasma cells are about 0.15-0.2 micrometers in diameter and 2.0-15.0 micrometers in length. Corn Stunt Spiroplasma cells are distributed only in the phloem tissue of the diseased plant including phloem of leaves, stems, roots, ears and tassels. Several leafhopper species transmit the spiroplasma, but the principle vector is the corn leafhopper Dalbulus maidis. Corn Stunt Spiroplasma is the causal agent of Rio Grande corn stunt, an especially severe disease of maize in Mexico and Central and South American countries. In the USA, Corn Stunt Spiroplasma has been found in many of the southern states and California. Corn Stunt Spiroplasma only infects maize and other relative teosintes and is more common in sweet corn and popcorn than in field corn. Symptoms on corn may vary with the corn cultivar. Young leaves are chlorotic or exhibit broad chlorotic striping and may purple as they age. Internodes are shorter and kernel set is reduced or absent. Symptom severity may partially depend upon when the plants were infected.

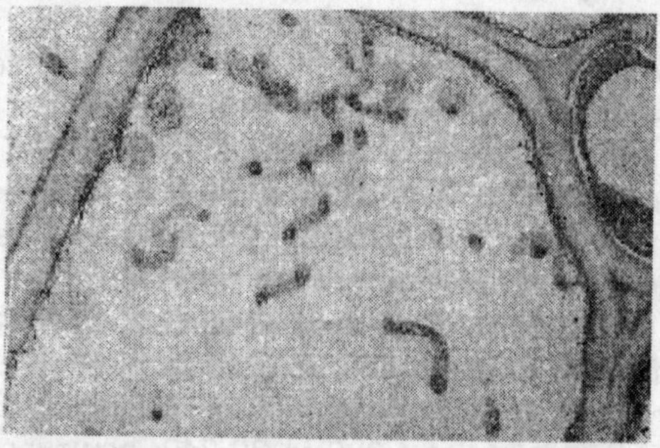

Fig. 5.1. Helical cells of Spiroplasma kunkelii in sieve cell in phloem tissue of a corn stunt diseased corn plant.

The use of ELISA is a reliable means of detection for Corn Stunt Spiroplasma. This ELISA test system is based on a polyclonal antibody for both capture and detection. The test is specific and relatively sensitive, producing a moderately strong reaction. Corn Stunt Spiroplasma can be detected easily in infected plant extract diluted at 1:270-810.This Corn Stunt Spiroplasma test shows no cross-reaction with healthy maize tissue. Also, no cross-reaction was found with other related corn pathogens such as maize rayado fino virus and maize bushy stunt phytoplasma. This test is recommended for testing Corn Stunt Spiroplasma in maize. It is suggested that the plant tissues be ground thoroughly with a sample buffer in order to release maximum Corn Stunt Spiroplasma cells from the phloem.

The mollicute Corn Stunt Spiroplasma (*Spiroplasma kunkelii*) is a leafhopper-transmitted pathogen of maize. Sequencing of the approximately 1.6-Mb genome of *Spiroplasma kunkelii* was initiated to aid understanding the genetic basis of spiroplasma interactions with their plant and leafhopper hosts. In total, 144712 nucleotides of non-redundant, high-quality Spiroplasma kunkelii genome sequence were obtained Sequence tags were searched against the Mycoplasmataceae and Bacillus/Clostridium databases. Results showed that, in addition to spiroplasma phage SpV1 DNA insertions, spiroplasma genomes harbor more purine and amino acid biosynthesis, transcription regulation, cell envelope and DNA transport/binding genes than Mycoplasmataceae genomes. This investigation demonstrates that survey sequencing is an efficient procedure for gene discovery and genome characterization.

To determine whether homologous recombination could be used to inactivate selected genes in *Spiroplasma citri*, plasmid constructs were designed to disrupt the motility gene scm1. An internal scm1 gene fragment was inserted into plasmid pKT1, which replicates in *Escherichia coli* but not in *Spiroplasma citri*, and into the *Spiroplasma citri* oriC plasmid pBOT1, which replicates in spiroplasma cells as well as in *E. coli*. Electro-transformation of *Spiroplasma citri* with the nonreplicative, recombinant plasmid pKTM1 yielded no transformants. In contrast, spiroplasmal transformants were obtained with the replicative, pBOT1-derived plasmid pCJ32. During passaging of the transformants, the plasmid was found to integrate into the chromosome by homologous recombination either at the oriC region or at the scm1 gene. In the latter case, plasmid integration by a single crossover between the scm1 gene fragment carried by the plasmid and the full-length scm1 gene carried by the chromosome led to a nonmotile phenotype. Transmission of the scm1-disrupted mutant to periwinkle (*Catharanthus roseus*) plants through injection into the leafhopper vector (*Circulifer haematoceps*) showed that the motility mutant multiplied in the insects and was efficiently transmitted to plants, in which it induced symptoms similarly to the wild-type *S. citri* strain. These results suggest that the spiroplasmal motility may not be essential for pathogenicity and that, more broadly, the *Spiroplasma citri* oriC plasmids can be considered promising tools for specific gene disruption by promoting homologous recombination in Spiroplasma citri, a mollicute which probably lacks a functional RecA protein.

General Diversity

All members of the genus Spiroplasma are obligately associated with insects (either as commensals, pathogens or mutualists) and may be one of the largest genera as a result of their diversification and co-evolution with their arthropod hosts. The Spiroplasmas display a combination of nutritional and microhabitat specializations and responses to ecological factors, which are important determinants of diversity. Spiroplasmas residing in arthropods are transmitted to plant surfaces by defecation and/or regurgitation of fluids on plant surfaces, or are introduced into the phloem by plant-sucking insects. Initial attempts to determine spiroplasma diversity relied on the estimate of approximately 30 million species of insects on the Earth and that one new cultivable and one noncultivable spiroplasma species was being identified in every 10 species of holometabolous insects examined. In essence these values would suggest that Spiroplasmas might eventually be the most speciose bacterial genus on the planet. Attempts were made to estimate taxonomic diversity of Spiroplasmas based upon the observation that one new Spiroplasma was found for every 10 species of insects examined in a large survey of insect families. Since there are an estimated 10 million insect species this quickly becomes an incredibly speciose microbial genera. However, although they well may end up being the most speciose microbial genus, more recent isolations have indicated that the rate of novel isolates will decrease as more and more Spiroplasmas sp. are identified and characterized. Recent results from collection sites on a North American latitudinal gradient from Nova Scotia to the Tropics suggest that the percentage carriage in insects varies geographically, with a probably peak of diversity in the Tropics. This general diversity has already been observed with spiroplasmas associated with the Tabanid (*Diptera*: Tabanidae) flies.

The Tabanidae have been a rich source of Spiroplasma species isolations primarily from horseflies and deerflies and arguably the most well studied of the spiroplasma insect hosts. Recent isolations from the tropics (e.g. Costa Rican highlands) have provided another rich source of Spiroplasmas. In a typical sampling, 12 isolations were made from 13 attempts from a single species of horse fly, which had not been previously sampled. Initial serological analyses indicated that these isolates did not belong to the same group. A single tabanid fly host was also observed on several occasions to be simultaneously infected with two or more Spiroplasma species.

Comparative collection studies from diverse geographical locations have provided a rich dataset to look at limitations on specificity. Spiroplasma isolates were obtained from tabanids collected on coastal plains from Florida to Nova Scotia, in the Rocky Mountains from Montana to New Mexico, Vermont, Texas and extensive collections in Bulloch County, Georgia. More than 400 tabanid isolates were analyzed using one-way serological deformation tests. One group of isolates (subgroup type strain DF-1-S. chrysopicola) was only obtained from deerflies (natural exclusive relationship with *Chrysops* spp. hosts) and had a wide range of distribution from Georgia to Wyoming. Two additional groups of Spiroplasma isolates had a southern, southeastern range as isolates were only found in deerflies collected in Georgia, Florida, and North Carolina. One of these two major groups (group type strain TG-1-S. gladiatoris) was comprised of over 100 isolates and was most commonly restricted to horseflies from 11 Tabanus species and was often found to occur in mixed infections with other spiroplasma species. The second major group of south-eastern restricted isolates (group type strain TAUS-1-S. tabanidicola) was restricted to a diversity of horsefly hosts. A fourth major group (group type strain TABS-2-S. helicoides) was also isolated from horseflies but distributed in the Midwest and southeastern collection sites. It is likely that extensive investigation of other insect hosts may also result in the same clustering of Spiroplasma isolates into related groups or subgroups.

Evolution

The majority of phylogenetic analyses conducted with members of the Mollicutes were based on the use of 16S rDNA sequence data. Using a large dataset and several of the available phylogenetic analysis programs, a very clear and consistent picture of Spiroplasma evolution has emerged, although there remain some areas of low resolution which is thought to be due to recently diverged and, therefore, very closely related strains.

Evolution of Spiroplasmas in Relation to other Major Groups of Eubacteria:

Although once considered to be primitive microbes evolving prior to the development of a cell wall, new phylogenetic data has provided a clear evolutionary path for this interesting group of microbes, which involves recent divergence and degenerate evolution from a eubacterial lineage and diversity arising from co-evolution with host organisms. Although many bacterial taxonomists recognize that complete bacterial genome sequences will be the gold standard for future phylogenetic analyses, the use of ribosomal DNA sequence analysis (primarily 16S rDNA but 5S rDNA is also used) has become the definite tool of choice for microbial phylogeny until such time as more complete genome sequences become available. Previous studies showed the utility of this gene sequence to be optimal for all levels of classification above the level of species and even at the species levels 16S rDNA sequence analysis serves to distinguish most species. Fortunately, the use of 16S rDNA sequence data was able to differentiate between species at the group level in spiroplasmas. However, a less highly conserved genomic region will be required for closely related subgroups.

Clostridium Lineage

Molecular data indicates that the Mollicutes arose mono-phyletically from a gram-positive, low G + C content Clostridial lineage of the Eubacteria (*Clostridium ramosum* and *C. innocuum*). The split is hypothesized to have occurred 600-800 mya with Spiroplasmas specifically emerging 300-600 mya. This would coincide with the evolution of arthropods (approximately 570 mya) providing a means for diversity of Spiroplasmas to be coupled with major arthropod host radiations.

Mollicute Lineage

Studies involving biochemical pathways, metabolism and rRNA gene homologies suggest the Mollicute evolution proceeded first with a divergence of the *Acholeplasmas* and *Anaeroplasmas*. The *Spiroplasma* branch split early from the *Acholeplasma* branch and the *Mycoplasma* and *Ureaplasma* branches appear to have evolved from the *Spiroplasma* branch after further genome reduction events. The plant-disease causing Phytoplasmas also evolved from the Acholeplasma branch. Thus each successive evolution was a result of degenerative (reductive) evolution so that the most recently evolved possess the smallest genomes.

Phylogeny of *Spiroplasma*

The phylogenetic analysis of the genus *Spiroplasma* was investigated using the 16S rDNA sequence analysis of 36 species, subgroups and strains. Several representatives from the genera *Mycoplasma*, *Mesoplasma*, *Entomoplasma* which were available on GenBank were combined with several single members from the genera *Ureaplasma*, *Phytoplasma*, *Acholeplasma*, *Anaeroplasma*, *Asteroplasma*, *Clostridium*, *Bacillus* and *Escherichia* to establish a dataset to clearly provide outgroup support for the monophylly of the *Spiroplasma* clade and to show the close relationship between the *Spiroplasmas* and members of the *Mycoplasma mycoides-Entomoplasma-Mesoplasma* cluster. Multiple phylogenetic analyses were performed on this dataset including neighbor joining, distance, maximum likelihood, and parsimony analysis. These analyses make significant changes to, and greatly improve, a preliminary UPGMA tree. The analysis revealed four main clusters that were monophyletic in all of the analyses. The representative tree indicated is that performed using Distance analysis.

Following the pioneering work of Woese, rDNA sequencing became the tool of choice for microbial phylogeny. Determination of 16S rDNA sequences has now been used in many attempts to resolve bacterial relationships at different levels. Results from these studies have established the bounds of resolution of the test. Charellier et al. and Boddinhaus et al. used this tool to examine serovars. Another intraspecific application was that of Whiley et al. on *Streptococcus* species. Bennasar et. al. examined *Pseudomonas genomovars*, and Yoon et al. examined intraspecific relationships among strains of *Nocardioides.* Brown et al. examined various *Mycoplasma* species for their placement within the genus. Takeushi et al. studied 23 serovars of *Erysipelothrix*. Studies of Mollicutes have permitted proper generic placements of certain strains. A noncultivable Spiroplasma associated with a sex ratio condition in a beetle was found to be basal on the tree and was closest to *Spiroplasma ixodetis*, a tick Spiroplasma. Various bacterial genera have received concentrated study by 16S technology. These include *Xanthomonas*, *Salmonella*, *Hyphomicrobium*, *Volcaniella*, *Clostridium*, *Prevotella*, *Nocardia*, *Acinetobacter* and

Chromobacterium. Koch et al. studied sequenced rDNA from seven bacterial genera. Mycoplasmas played a prominent role in application of the 16s rDNA technique to bacterial systematics. The technique has been applied at the family and/or intrafamily level for Pasteurellaceae. This molecule has also been used for ecological groups of bacteria, e.g., Riegel et al. Finally, the advent of 16S sequencing has enabled relationships among noncultivable organisms to be explored for the citrus greening bacterium; *Haemobartonella* and *Eperythrozoon* and *Phytoplasmas*. The results of these and many other studies have shown that the rDNA sequence studies are ideal at all levels of classification above species. Even at the species level, this method separates most species well, but, because it is so highly conserved, it may become less effective for discriminating among pairs of species that are closely related. The resolution of the method varies somewhat from genus to genus, but Stackebrandt and Goebel proposed a rule of thumb, based on data comparing the two techniques, that at homology levels of 16S rDNA-rDNA higher than 97%, the conventional test-DNA-DNA hybridization-should also be performed. A robust data base of 16S rDNA sequences exists for spiroplasmas. Weisburg et al. studied seven species and later developed PCR amplification methods to expedite future research. Dodge et al. recently presented preliminary trees for more than 40 Spiroplasma strains in a poster at the International Congress of the International Organization of Mycoplasmology. These sequences are now being analyzed by current phylogenetic methods. A final word of caution about 16S rDNA sequence analysis is warranted. Sequence heterogeneity poses some problems in analyses. Also, sequences may be highly conserved in some cases, obscuring species identities.

The three major techniques used in mollicute systematics have different resolving powers. Although 16S rDNA sequence analysis has the broadest relevance across higher taxa, including the genus level, its power at the species level is still uncertain. DNA-DNA hybridization, although cumbersome, has excellent resolving power at the group (species) and subgroup levels. Serology also has excellent resolving power at the group and subgroup levels but may be inadequate to discriminate at the fine intraspecific level. It is this level that poses the greatest challenge in future systematic studies of microorganisms.

Spiroplasma characteristics such as motility, chemotaxis, adherence to host cells, small size, high surface to volume ratio (for transfer of nutrients into the cell), cell wall-lessness, low G + C content of DNA (hypothesized to accumulate in host-dependent organisms), an evolutionary rate hypothesized to be higher than normal mutation rates and reduced biosynthetic capabilities correlate well with life in host microhabitats that are transient, diverse and nutritionally rich. The spiroplasmas probably arose as gut symbionts in arthropods and later evolved to exploit emerging habitats within insects, ticks and animals. Radiation of the genus Spiroplasma (340 mya) coincides with the emergence of lineages of hemipteran (leafhopper) and insects that go through a complete metamorphosis. All of this information provides a clear justification for understanding microbial host range and phylogeny when determining the evolutionary course of host-dependent microbes.

The use of comparative metabolism has also recently been explored as an additional means to determine the phylogenetic relationships among the different Mollicute genera. DNA sequence data alone may not definitively determine function. Many factors influence protein availability and function including: peptide folding (tertiary structure); peptide-peptide interactions (quarternary structure); substrate and cofactor concentrations; and, post-translational modifications. These factors may cause the genotype and the phenotype to actually differ in an organism and must be considered when looking at the complex question of evolution.

Although microbiologists are more or less convinced that 16S rDNA sequencing yields reliable results, they are not prepared to base classifications solely on any single character. Rather, they prefer a "total evidence," or polyphasic, approach that utilizes all the information available about a taxon. Polyphasic taxonomy utilizes different kinds of information (phenotypic, genotypic, and phylogenetic) and is essentially a consensus method that is coming into increased usage. Types of information used in bacterial polyphasic taxonomy include genotypic (rDNA-largely 16S-sequence homology, DNA-DNA hybridization, G+C content, DNA-based typing methods such as RFLP, plasmid analysis, codon usage, and PCR-based techniques) and phenotypic methods (morphological, physiological, and biochemical features), which may include cell wall composition, whole-cell protein analysis, and serology. Many of these techniques have been applied in taxonomy of the Entomoplasmatales, and development of a polyphasic classification of the genera of this order will no doubt emerge in the not-too-distant future.

Since its designation as a separate genus some 30 years ago, Spiroplasmas have been well documented in a wide range of hosts and as the causative agent of several plant and insect diseases. One major area of research is the continued identification and taxonomical characterization of new *Spiroplasma* sp. combined with a determination of phylogenetic relationships among the various *Spiroplasma* sp. and between the Spiroplasmas and other members of the Mollicutes and Eubacteria. Although most phylogenetic analyses have been dependent on 16S rDNA sequence data, progress in two *Spiroplasma* sp. genome sequencing projects will provide new genomic regions for comparative focus. The co-evolution of Spiroplasmas with their arthropod hosts has provided an additional research focus to study host specificity and attachment. The diversity of symbiotic relationships between Spiroplasmas and their hosts has led to the study of commensal, mutualistic, and pathogenic relationships. Pathogenesis in insect hosts or in plants, transferred by insect hosts, is a major research focus, which requires attachment and invasion into insect tissues beyond the initial infection site, and successful movement to other tissues.

Descriptions of *Spiroplasma citri* cluster

16S rDNA-based phylogenetic analysis of the spiroplasmas results in several clearly defined and differentiated clusters of groups and/or subgroups. Three major clades are differentiated within the *S. citri* cluster in all phylogenetic analyses.

Group I and II (citri) clade

The Group I clade is tightly clustered with the representative type strain from Group II. This particular clade is of great interest as it has a wide diversity of host associations and contains a wide variety of spiroplasma pathogens. Members within this cluster are specialists in ticks, honeybees, leafhoppers (plant phloem sucking insects), plant/flower surfaces and Drosophila (sex-ratio organism). Geographical distribution ranges from the New World tropics, Mediterranean regions of Europe, Africa, U.S. (warm and cold temperate regions), and Asia. The group I Spiroplasmas form a very diverse cluster that is divided into eight subgroups.

Group VIII (chrysopicola) clade

The group VIII cluster is a large taxon of spiroplasmas primarily isolated from tabanid flies. Member of this group are physically the smallest of Spiroplasmas-passing through 220 nm filters. This group also has a G + C content (28-31%) on the high end of the spiroplasma range and all members has the ability to utilize arginine. Strains were isolated from a diverse geographic range including the U.S., Europe, Costa Rica and Australia. Three formal subgroups were proposed for group VIII strains on the basis of DNA-DNA hybridization and serological analyses. The type strains for the subgroups are: Subgroup VIII-1 *Spiroplasma syrphidicola*; subgroup VIII-2 S. chrysopicola; and subgroup VIII-3 S. sp. strain TAAS-1. Several recent isolations and preliminary serological screening indicates that there are several other candidates for group VIII subgroup status. Unlike spiroplasmas isolated from horseflies in the apis group (below), Group VIII spiroplasmas are short in length, have higher G + C contents of their DNA (29-30% vs. 26-28%), and, unlike long apis clade spiroplasmas, they utilize arginine.

Spiroplasma mirum clade

This cluster contains the single species *S. mirum* (Group V), from a tick. This species is always basal to the Groups I, II and VIII supercluster in all phylogenetic analyses. *S. mirum* is the most basal species that shows the typical spiroplasma helicity and motility consistently.

Descriptions of *Spiroplasma apis* cluster

The majority of the remaining *Spiroplasma* spp. form a large cluster that is sister to the *Spiroplasma citri* cluster described previously. This group can be separated into nine clades which, when combined, are loosely referred to as the *S. apis* clade. Clades are grouped together in some cases by common insect hosts or plant feeding sites and insects known to feed at those sites however in other cases the diversity of spiroplasma insect hosts does not seem to make a cohesive unit but further study of spiroplasma transfer flow within an ecosystem and an increase in the number of isolates in each group may clarify the reason for the close evolutionary relationships.

Spiroplasma lampyridicola clade

The *S. lampyridicola* clade consists of two members, *S. leptinotarsae* and *S. lampyridicola* which are both beetle specialists associated with the Colorado Potato Beetle and firefly beetles, respectively. *S. leptinotarsae* is transmitted from beetle to beetle on leaf surfaces. *S. lampyridicola* was isolated from both larvae and adults, but the transmission mechanism is unknown.

***Spiroplasma sabaudiense* clade**

This clade consists of groups XIII, XXVI, and XXIX from a diverse range of insects including mosquito (postulated to feed on flowers) and flower-visiting tiphiid wasp isolates as well as a scorpionfly gut isolate. The scorpionfly isolate is questionable as this is a predatory insect so it is unknown if this was a transient spiroplasma infection or a normal gut constituent. The common feeding site for the other groups can explain the transfer and close association of these groups.

***Spiroplasma litorale* clade**

This clade consists of horsefly isolates with representatives from groups XIV, XVII and XVIII-all have a G + C content in the 25/26 range and all are non-arginine catabolizers. Isolates linked to group XIV were also isolated from beetles, which has raised the interesting issue of transmission and maintenance of the spiroplasma culture during winter. It is hypothesized that perhaps the beetles provide a reservoir for spiroplasmas to overwinter and that after the beetles feed on common flower sources with the flies in the spring the transmission is then accomplished to the fly hosts.

***Spiroplasma helicoides* clade**

The *S. helicoides* clade consists of three more tabanid groups (groups XXIII, XXXII, and XXXIV) specifically isolated from horseflies.

***Spiroplasma floricola* clade**

This clade contains groups III and XII, which are either isolates from flower surfaces or from cantharid beetles, which are frequent visitors of flowers.

***Spiroplasma monobiae* clade**

Members of groups VII, XVI-3, XVI-2, and XXV are represented in this clade and also come from a diverse group of insect hosts including beetles and wasps and an isolate from a flower surface. It should be noted again that the insect hosts all feed on plant surfaces. The third subgroup of XVI (XVI-1) was not sequenced for this survey but the close DNA-DNA hybridization data indicating a close association of the subgroup 1 isolate with that of subgroups 2 and 3 is probably indicative of subgroup 1 also belonging to this clade, but that has yet to be determined definitively.

Spiroplasma sp. strain W115 clade

Both members of this clade, strain W115 (group XXI) and strain BIUS-1 (group XXX) are very closely related phylogenetically and both were isolated from flower surfaces so the actual insect host is unknown for both, but it can be assumed they are probably insects that feed on flowers at some point in their life cycles.

Spiroplasma culicicola clade

Representatives from groups X, XI, and XXIV were observed in this cluster with representative spiroplasma isolates from a mosquito from the French Alps, a wasp from Maryland and a flower surface in China. The ecological or evolutionary ties between these isolates, which are diverse from the perspective of both geographical isolation and insect host, are unclear.

Spiroplasma montanense clade

The *S. montanense* clade consists of four groups (XXXI, XXII, IV, and IX), which were isolated from horseflies, honeybees, Green June Beetles and mosquitoes. The close relationship between groups IV and XXXI is explained by the isolation from horseflies in both groups although the spiroplasma type strain for group IV is isolated from a honeybee.

Descriptions of *Mycoplasma mycoides* cluster

Mycoplasma mycoides /Mesoplasma/ Entomoplasma clade: Many previous phylogenetic studies have placed the type strain for the genus *Mycoplasma* with a subgroup of other *Mycoplasma* sp. along with members of the genera *Mesoplasma* and Entomoplasma. All of these form a cluster that appears most closely related to the large *S. apis* clade. Although the *Mycoplasma* spp. are tightly clustered together, the *Mesoplasma* and Entomoplasma do not form two coherent clusters, but are intermixed in one paraphyletic group. Although the two can be differentiated by growth requirements: Entomoplasmas requiring serum and Mesoplasmas being able to grow in the presence of Tween 80 instead of serum, it has been suggested that the two be combined into a single genus.

Descriptions of *Spiroplasma ixodetis* cluster

Spiroplasma ixodetis clade: Originally this cluster was thought to consist of a single species, *Spiroplasma ixodetis* (Group VI), which always appears at the root of the Spiroplasma clade regardless of the type of phylogenetic analysis used. However, a recent isolation frc m a pea aphid has shown that this cluster now contains several strains. The *S. ixodetis* cluster appears phenotypically to be the transitional species prior to Spiroplasma speciation. Some members of this group exhibit the classical spiroplasma helicity, while others are filamentous. The genome is 2,220 kbp in size, which is the largest for the Spiroplasma genus, and just 500 kbp more than the genome of *A. laidlawii*. The evolutionary distance of *S. ixodetis* from other *Spiroplasma* spp. is significant.

Pathogenesis

Spiroplasmas established a vast array of interactions with plants and insects, ranging from epiphytic to symbiotic to pathogenic. Some Spiroplasma spp. does not replicate in insects and are found on flower surfaces or insect intestines. Other *Spiroplasma* spp. replicate in insects, thereby causing disease. Three Spiroplasma spp. are transmitted to plants by leafhoppers and replicate in the phloem tissue of plants. A fourth type of spiroplasma is transmitted to next generation insects through the embryo and kill the male progeny in the female insect, and, therefore are called sex ration organisms (SRO). At last, *Spiroplasma mirum* isolated from rabbit ticks is shown to infect rabbits when artificially injected, however the infection has never occurred in nature. Though mollicutes are common as pathogens of plants, they are found even more frequently in insects and other arthropods. Some spiroplasmas cause disease in their arthropod hosts, but most insects and mollicutes seem to share a symbiotic or at least mutually non-detrimental co-existence. Such relationships between plant pathogens and the insects that serve as their vectors may reflect a long period of co-evolution of the two species. The physical and physiological factors that determine whether a nonphytopathogenic spiroplasma is able to penetrate the insect's intestine wall and colonize the hemolymph may also determine whether phytopathogenic mollicutes are transmissible by a certain insect.

The three plant pathogenic *Spiroplasma* spp., *S. kunkelii* (corn stunt spiroplasma or CSS), *S. citri* (citrus stubborn spiroplasma), and *S. phoeniceum*, are transmitted by leafhoppers from plant to plant. In insects, spiroplasmas are likely to adhere to receptors on the apical plasmalemma of the leafhopper gut and are absorbed by gut epithelial cells. After migrating through the cell, they are released into the space between the basal plasmalemma and the basal lamina and, from there, move to the hemolymph, where they circulate and replicate. Spiroplasmas move from the hemolymph into the salivary ducts and saliva through cell layers of the salivary glands. With the saliva, spiroplasmas are introduced into the phloem of plant by feeding leafhoppers. The latent period in the insect is at least two weeks.

Fig. 5.2. A photograph showing Dalbulus maidis, the most important vector of Spiroplasma kunkelii, feeding on maize.

The interactions of *S. kunkelii* with different leafhoppers of the genus Dalbulus vary. In efficient vector systems, infections with these mollicutes improve leafhopper fitness, whereas in inefficient vector systems, infections result in virulent pathogenicity. For example, infection of *D. maidis* by *S. kunkelii* enhances the resistance to cold temperatures of the leafhopper and increases its ability to survival in maize free winter periods, thereby, benefiting both the plant pathogen and the insect. The differences in association can be in part explained ecologically; more ancient associations of spiroplasmas with their host organisms have less detrimental effects than relatively recent ones.

In order for the spiroplasmas to maintain themselves in their host organisms (regardless of strict or general specificity) they must have the ability to bind to the insect epithelial cells to avoid being flushed through the digestive tract. In many cases this is the extent of attachment, but in order for an organism to become pathogenic, the spiroplasma must be able to invade into the hemolymph. In some extreme cases, for example in the leafhopper hosts which transmit plant disease, the spiroplasma must also pass into other tissues to specifically reach, in the case of *S.*

citri, the salivary glands of their leafhopper hosts. In the case of the sex-ratio spiroplasma, the microbe must be able to specifically attach and invade into the ovaries. Some transmission cycles appear to be very complex, however little is known about even the most basic attachment mechanism. Much work has been done in the Mycoplasmas, which have given some clues to spiroplasmologists but no definitive or homologous answers have yet been elucidated.

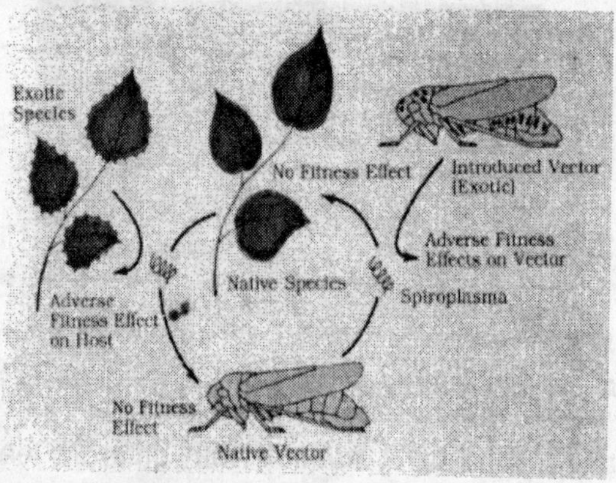

Fig. 5.3. Schematic representation of the interaction of spiroplasmas, leafhoppers and plants. Spiroplasmas do not affect adapted host plants and insect vectors but cause adverse effects on introduced or foreign.

All of the initial studies involved in the question of attachment and pathogenicity have focused in the arena of plant pathogens. Spiroplasmas exhibit different types of interactions with their insect and plant hosts ranging from pathogenic to mutualist relationships. Some Spiroplasma spp. does not replicate in insect cells and are found on flower surfaces or within the insect gut. Other *Spiroplasma* spp. can replicate in insects, usually after invading into the hemolymph and other tissues, thereby causing disease in their insect hosts. Thus far, pathogenic spiroplasmas tend to be limited to plants, beneficial insects and limited transformation of vertebrate cells or immunocompromised vertebrates.

Transmission

Transmission of spiroplasmas to plants occurs when leafhoppers feed from plant phloem. Bacteria are introduced into the phloem with insect saliva. In plants, spiroplasmas remain restricted to phloem tissues and systemically spread throughout the plant. Spiroplasmas cause severe symptoms, such as, stunting, yellowing of leaves, and yield losses in a variety of economically important crops. Insect-transmitted plant-pathogenic spiroplasmas are not transmitted via plant seeds, and are not transovarially transmitted to next-generation leafhoppers. Therefore, they cannot survive outside their hosts and are dependent upon transmission for survival and dispersal. The molecular and genetic mechanisms underlying the diverse interactions between plant pathogenic spiroplasmas and their plant or insect hosts remain unknown.

Host Range

Discovery of the arthropod, plant, and flower surface habitats of spiroplasmas has led to an explosion of described isolates and new Spiroplasma species in the last 20 years. By far the most common hosts for spiroplasmas are the arthropods. Spiroplasmas are found most commonly in the guts of insects as part of the normal gut flora, but some pathogens can enter the hemocoel and other organs such as the salivary glands, ovaries and brain cells. Initial isolation attempts of Spiroplasmas sp. from their insect hosts may be difficult, as many new isolates often require a transitional period with insect cell co-culture or preliminary anaerobic isolation prior to adaptation to artificial media under aerobic conditions. Thus far, Spiroplasmas have predominantly been isolated from the six most evolutionarily advanced orders of insects although one Spiroplasma sp., strain PALS-1, was isolated from a dragonfly (*Pachydiplax longipennis*) which belongs to a more ancient order of insects (*Odonata*). However, since the dragonfly is predaceous on other insects, the isolate might have been transmitted during a feeding. The six main orders of insects serving as *Spiroplasma* sp. hosts include: 1) Hymenoptera with spiroplasma isolates from honeybees, bees, and vespid wasps; 2) Coleoptera with spiroplasmas isolated from a variety of beetles including the green June beetle, soldier beetle, cucumber beetle, Colorado Potato beetle, scarabaeid beetle, firefly beetles; 3) Diptera with spiroplasmas isolated from syrphid flies, mosquitoes, tabanids, and fruit flies; 4) Lepidoptera with isolates from butterflies and moths (in both larval and adult stages); 5) Homoptera with Spiroplasmas isolated from leafhoppers; and, 6) Hemiptera with isolates from the green leaf bug.

Some recent isolation has increased information concerning host ranges and diversity. For example, an isolate from a Satin Moth caterpillar has now provided evidence for spiroplasma infection during all Lepidopteran life stages including eggs, larva at different instars, pupae and adults. The first isolation from a midge insect host has also been reported increasing the diversity of hosts in the order Diptera. A male-killing isolate (sex ratio organism) was obtained from a ladybird insect (*Adalia bipunctata*) and identified as a *Spiroplasma* sp.. Ticks also serve as a host for several spiroplasma strains. These include rabbit ticks from which Spiroplasma strains SMCA and 277F and the Ixodid tick which serves as the host for Spiroplasma strain Y32.

Although Spiroplasmas are always associated with an insect or tick host at some point, several primary isolations have also been made from plants (both internally and from the surface). Spiroplasmas were isolated from both the surfaces of flowers and the sap in the sieve tube elements of phloem. As many insects feed primarily on nectar it has been suggested that the Spiroplasmas found on flower surfaces were deposited there by feeding insects and thus serves as an excellent agent for transmission of the bacteria from one host to another. Those isolates from the sieve tube elements were deposited there by plant-sucking hemipterans (e.g. leafhoppers).

Adaptation

The close associations between Spiroplasmas and their hosts has led to selective pressures that have optimized some Spiroplasmas for strict specificity in their hosts and provided others with the ability to infect a range of hosts. This close dependence of the Spiroplasmas on a host provides an excellent model system to study co-evolution.

Host cell specificity

The diversity of characteristics among spiroplasma strains carries over into host(s) interactions. Although only a few studies have been conducted to experimentally explore or environmentally survey host specificity in spiroplasma associations, both strict specificity and group specificity has been observed.

Strict specificity

With the information available to date, only a few spiroplasmas have been identified as being host specific. A classic example of strict host specificity is the group XX spiroplasma isolate from a Colorado potato beetle, which was found to be specific to its host. The Colorado potato beetle isolate, *S. leptinotarsae*, did not persist when fed or inoculated into other insects including other beetles.

General specificity

The most well studied group of spiroplasma insect hosts, the Tabanidae, also affords the best evidence for general host specificity by some Spiroplasmas. Natural isolations and laboratory infections afford a look at broader implications of non-specific host interactions. At issue is an attempt to understand the natural route of spiroplasma colonization of their tabanid hosts, especially when tabanid larvae are rarely infected with Spiroplasmas. The firefly was suggested as a potential alternate host as they are also a rich source of spiroplasmas. Attempts were made to simulate natural predation conditions in the laboratory whereby tenebrionid pupae were artificially infected with tabanid Spiroplasmas and used as a food source for fireflies. However, no infection of the firefly larvae by predation on infected mealworm beetle pupae (*Tenebrio molitor*) was observed. An alternate suggestion for transmission of Spiroplasmas might be due to common carbohydrate feeding sites (e.g. nectarines and honeydew deposits) among adult fireflies and tabanids. Evidence supporting this hypothesis was reported from findings that tabanids could become readily infected with Spiroplasmas by ingestions of Spiroplasmas placed in 5% sucrose solutions. This infection through common feeding sites may explain the horizontal transmission of Spiroplasmas from one infected tabanid to another and also account for the higher percentage of mixed or multiple spiroplasma infections observed in tabanid hosts.

Another example of generalized specificity is exhibited in laboratory experiments with the group XII spiroplasma originally isolated from the corn rootworm beetle. Naturally this spiroplasma can invade into the hemolymph of its corn rootworm beetle host. In laboratory feeding experiments, the group XII spiroplasma was also able to invade into the hemolymph of the Colorado potato beetle.

Diseases

In most cases, the ability of a spiroplasma to be pathogenic in its insect host is a direct result of the adaptation to invade into the hemolymph from the initial site of infection, attachment to the gut epithelial cells. A clear example of host insect pathogenicity was observed in Honeybees with two different spiroplasma strains. Honeybee mortality is clearly linked with the hemolymph invasion by *S. melliferum*. Similar pathogenicity was observed with *S. apis*. Male lethality was also observed by those spiroplasmas, which have the ability to invade ovaries of specific host organisms. There is also a negative impact on survival and fecundity of the Dalbulus leafhoppers carrying *S. kunkelii* although further studies are needed confirm this pathogenicity and to determine if other leafhopper hosts are similarly impacted.

Once in the plants, Spiroplasmas remain restricted to the phloem tissue where they spread throughout the plant. Spiroplasmas cause severe symptoms, such as, stunting, yellowing of leaves, and yield losses in a variety of economically important crops. These organisms are not transmitted via plant seeds, and are not transovarially transmitted to next-generation leafhoppers. Therefore, they cannot survive outside their hosts and are dependent upon transmission for survival and dispersal. The molecular and genetic mechanisms underlying the diverse interactions between plant pathogenic spiroplasmas and their plant or insect hosts remain unknown.

Plant physiologists have been studying the Spiroplasma-induced infections intensively in the past few years in an attempt to describe the infection process more fully. The laboratory of Dr. Jacqui Fletcher (Oklahoma State University) has focused on the three lines of S. citri BR3, which have different degrees of transmissibility by the beet leafhopper host (*Circulifer tenellus*). The three different derivatives of strain BR3 used in pathogenesis studies are: 1) BR3-3X is the original triply cloned isolate from horseradish plants with brittle root disease and has been maintained for over 10-years; 2) BR3-T which can be transmitted from plant to plant by its leafhopper vector (*C. tenellus*) was reisolated from BR3-3X infected plants after repeated transmission; and, 3) BR3-G which can only be maintained in plants through grafting and has lost its ability to be transmitted by leafhopper. Physical genome mapping of all three isolates indicated significant changes in chromosomal DNA restriction patterns due to inversion and deletion events.

In vitro adhesion assays to identify proteins that may be involved in the transmission of Spiroplasmas from their insect hosts to plants indicate that the adhesion-related protein in *S. citri* may play a critical role in the passage of the pathogen through the insect host (through the intestinal wall to the salivary glands). Development and use of an established cell line of the leafhopper Circulifer tenellus to characterize Spiroplasma citri-vector interactions. Electron microscopic evaluation of *S. citri* 's traversal of the gut epithelium and salivary glands of the leafhopper vector, *C. tenellus*, was made. Mechanisms by which *S. citri* enters and exits *C. tenellus* gut epithelia, salivary gland, and cultured cells were suggested by spiroplasma location and the nature and position of host cell constituents. The inhibition by proteases of *S. citri* adhesion to cells of the vector leafhopper *C. tenellus* was accompanied by loss of an 89 kDa protein and the appearance of a new band of about 46 kDa. Both adherence ability and the 89 kDa protein were restored after spiroplasma growth. The 46 kDa protein reacted with an anti-P89 polyclonal antibody, suggesting that it is a breakdown product of the 89 kDa protein. These findings suggest a role for P89 in spiroplasma-vector cell binding.

Dr. Joel Renaudin (INRA-Bordeaux) is using transposon mutagenesis to create motility and nonpathogenic mutants of *S. citri*. Two mutants of Spiroplasma citri were studied in detail. One, GMT553, does not induce symptoms in plants and is unable to use fructose as a result of integration of Tn4001 into the fructose operon. The other, G540, is a non-motile mutant. In both cases complementation and gene disruption experiments were successful. The gene involved in motility, scm1, codes for a protein of 409 amino acids with no significant homology in databanks. Complementation experiments were able to restore the motility mechanism in these mutants providing clear evidence of the association of the scm1 gene product in the *S. citri* motility. This protein is mainly hydrophobic with a eubacterial signal peptide and several transmembrane motifs. Disruption of the gene and transmission to plants showed that scm1 is not required in symptom development. The possible role of this gene in insect transmission is under investigation to determine if motility may be linked to the process of pathogenesis in Spiroplasmas.

To determine the role of the *S. citri* scm1 gene, homologous recombination with a replicative, pBOT1-derived plasmid carrying an internal fragment of the scm1 gene was used to disrupt the scm1 gene. Transmission of the scm1-disrupted motility mutant to periwinkle plants through injection into the leafhopper vector (*Circulifer haematoceps*) as well as by feeding the insects on motility mutant-infected plants showed that the motility mutant was able to cross the gut epithelium, multiply in the insects, and was efficiently transmitted to plants in which it induced symptoms similar to the wild-type *S. citri* strain. These results suggest that *S. citri* motility, and particularly the rotational movement that is lost in the scm1 mutant, may not be essential for insect transmissibility and pathogenicity to plants.

The pathogenic role of fructose uptake by the spiroplasma was studied in more detail with additional mutants obtained by gene disruption and by selection on xylitol, a toxic fructose analog. The non-phytopathogenic mutant GMT 553 was obtained by insertion of Tn4001 into the first gene of the fructose operon, the putative regulatory fruR gene, the other genes being the permease fruA gene and the 1-phosphofructokinase fruK gene. Complementation of the fructose- mutant GMT 553 with fruA, fruA+fruK or fruR+fruA+fruK restored fructose utilization, suggesting a relationship between fructose utilization and pathogenicity. Additional fructose- mutants were produced either by fructose operon disruption or selection of spontaneous xylitol-resistant strains. These various mutants, including GMT 553 and complemented GMT 553 strains, were transmitted via leafhopper-vectors to periwinkle plants. Symptom development, spiroplasmas titers in plants, and reversion of fructose- mutants to fructose+ revertants in the plants were studied. The results show that early and severe symptoms are obtained with spiroplasma strains able to use glucose and fructose, but late and mild symptoms with strains able to use glucose but not fructose. Fructose utilization by the spiroplasmas could impair sucrose loading into the sieves tubes and results in accumulation of carbohydrates in source leaves and depletion of carbon sources in sink tissues. Dr. Monique Garnier (INRA-Bordeaux) explored the use of differential mRNA display for *S. citri* and periwinkle disease in order to characterize plant genes. Four different genes were selected as being up regulated or down regulated in response to infection. A search of the GenBank database for protein homology indicated one of these genes was similar to Oryzae sativa transketolase and another homologous to delta-C-methyltransferase.

Unusual Sex-Ratio Disorders

A third type of pathogenic spiroplasma is transmitted to next generation insects through the embryo (transovarial transmission) and kills the male progeny in the female insect. This group of organisms is therefore referred to as sex ratio organisms (SROs). Insects with antagonistic sibling interactions (e.g. male death benefits female siblings) are those most impacted by the SROs. To date, SROs have been isolated from *Drosophila* spp., two different beetles, *Adalia bipunctata* and *Harmonia axyridis* and *Danaus chrysippus* butterflies. The isolate from *Drosophila willistoni* is designated *S. poulsonii* (Group II) and eliminated all male progeny of infected females. The Spiroplasmas isolated from the *A. bipunctata*, *H. axyridis* and *D. chrysippus* hosts were analyzed by rDNA sequence analysis and it was determined that they were both most closely related to *S. ixodetis* (Group VI). This information indicates that the male-killing trait evolved at least twice within the genus Spiroplasma.

Spiroplasma Infections of Animals

Only one *Spiroplasma* sp., *Spiroplasma mirum* strain SMCA (isolated from rabbit ticks), has been shown to infect animals. Early passage SMCA produced cataracts in suckling mice and rats when artificially infected in pathogenicity tests. The strain was also found to be pathogenic for 7-day-old chick embryos. It is important to note that these pathologies have never been observed in nature and that the animals that were challenged in these experiments were immunocompromised due to immaturity at the time of the challenge. *S. mirum* strain SMCA can also induce a malignant transformation of mammalian NIH-3T3 cells which produced tumors in nude and BALB/c mice. The transformed cells were microscopically shown to contain spiroplasma structure; DNA alone from strain SMCA was not sufficient to induce the cell transformation.

Spiroplasmas and Creutzfeldt-Jakob Disease and Scrapie

Recent studies indicate a correlation between Spiroplasma infections and transmissible spongiform encephalopathies (TSE) in humans and animals. Evidence was presented showing the association of *Spiroplasma* sp. with transmissible spongiform encephalopathies using Spiroplasma-specific 16S rDNA primers in the PCR amplification of infected brain tissue. Results indicated PCR amplification from 13 of 13 Creutzfeldt-Jakob diseased brain tissue samples from humans and 5 of 9 Scrapie diseased brain tissue samples from sheep. None of the 50 control brain tissue samples resulted in a PCR product. Direct sequence analysis of the amplified DNA confirmed the presence of spiroplasma-like DNA, which was 99% homologous to S. mirum. This information requires more study as it currently shows association but not causation of *Spiroplasma* sp. with the TSE diseases.

Host Interactions: Commensalistic Relationships

Most Spiroplasmas are currently classified as commensals in their arthropod hosts. These organisms attach to the epithelial cells of the host midgut and do not invade into the hemocoel or other tissues of the host insect.

Host Interactions: Mutualistic Relationships

Although the majority of symbiotic relationships that exist between Spiroplasmas and their hosts either fall into the category of pathogenic or commensalistic relationships, there is some evidence for the existence of mutualistic relationships as well. Plant-pathogenic mollicutes can be a mutually beneficial symbiont of leafhopper hosts. For example, the interactions of *S. kunkelii* with different leafhoppers of the genus Dalbulus vary. In efficient vector systems, infections with these mollicutes improve leafhopper fitness, whereas in inefficient vector systems, infections result in virulent pathogenicity. For example, the infection of *D. maidis* by *S. kunkelii* enhances the resistance to cold temperatures of the leafhopper and increases its ability to survive maize free winter periods, thereby, benefiting both the plant pathogen and the insect. The differences in association can be in part explained ecologically; more ancient associations of spiroplasmas with their host organisms have less detrimental effects than relatively recent ones.

Taxonomy

Spiroplasma taxonomy is grounded in the systematics of the class Mollicutes; in fact, the type species of the class Mollicutes is a member of the Spiroplasma clade. Recently, two of the standard characters of mollicute (and spiroplasma) higher level taxonomy-sterol requirement and genome size-have collapsed, in that they are variable within taxa. Three major criteria that remain useful are 16S rRNA sequence data, DNA-DNA reassociation data, and serology. The highest resolution is provided by 16S rRNA sequence data, which is useful for many or most species distinctions, and at all higher taxonomic levels. The resolving powers of DNA-DNA data and serology are similar; both are excellent at the species level, but neither are useful above the species level. The resolving powers of these techniques at the fine intraspecific level are uncertain; this aspect of mollicute systematics is a great challenge. Spiroplasma species are now defined by serology; in the case of partially related strains, DNA-DNA homology must be applied to determine whether candidate species have less than 70% reassociation with existing species. Today, 34 groups and 14 subgroups are recognized. Thirty-four of these species and subgroups have been accorded species status. Subgroups I-1, VIII-1, and XVI-1 are the representative taxa for their respective groups, so that there are 45 designated taxa in all. There have been four revisions of the original classification, the most recent in 1998.

The taxonomy of the genus Spiroplasma actually began in 1973, when Saglio et al. published a description of the cultivated filamentous mycoplasma-like agent isolated from tissues of citrus plants affected with stubborn disease. They adopted the trivial name proposed by Davis and Worley as the generic name and named the organism *Spiroplasma citri*. In 1976, the ICSB Subcommittee assembled a group of Spiroplasma workers at the IOM meeting in Glasgow. At this meeting, the task of evaluating various techniques (e.g., serology, DNA-DNA hybridization, G + C determination, electrophoresis) was partitioned and applied to a number of Spiroplasma strains then available. By 1980, results of the comparison were complete, and the team published a group classification of spiroplasmas. Five groups were proposed; group I was subdivided into three subgroups. The classification included a set of new strains from the honeybee and flower surfaces. The work of Clark and his colleagues, investigating insects as habitat for spiroplasma, then led to a plethora of new groups and species. In the Entomoplasmatales, the species concept was developed by determining DNA-DNA hybridization values among Spiroplasma taxa with varying degrees of serological cross-reactivity in Spiroplasma deformation and metabolism inhibition tests. Organisms with little or no cross-reactivity in either serological test showed little or no DNA-DNA homology. Groups of strains with a high degree of serological cross-reactivity showed high degrees of DNA-DNA homology (80-100%) and could be grouped together as a genomic species. Pairs of organisms that exhibited partial cross-reactivity in serological tests tended to have partial DNA-DNA relationships (20-80%). Although taxa with 80-90% homology with a type strain were eligible for subspecies designation, no subspecies have been designated in Entomoplasmatales. In the original classification paper of Junca et al., DNA-DNA homology tests were performed alongside the serological studies. From the results of these comparisons, it was inferred that serology could be used as a predictor of DNA-DNA homology values and could therefore be used to infer the status of emerging taxa as probable new genomic species. After this initial calibration of techniques, serology replaced the expensive and time-consuming methodology of DNA-DNA homology. A major factor in Spiroplasma taxonomy concerns cultivation of the organisms. Some organisms are sufficiently fastidious that they are unwieldy to manipulate at the bench; others are relatively fast growers. Approaches to cultivation in both defined and undefined media have been reviewed, and formulas for media have been provided. The two schools in Spiroplasma cultivation separated early, with the concurrent cultivation of the corn stunt spiroplasma in 1975 by two groups. Chen and Liao cultivated the organism in a very simple medium (C3G) and continued to explore means to simplify the media further, as summarized by Chang. Williamson and Whitcomb cultivated the agent in a complex medium, which was a compromise between current Spiroplasma media and a complex tissue culture medium. Eventually this medium, with very slight modification, became M1D, which has been used for the majority of cultivated spiroplasmas. In 1977, Jones et al. attempted to dissect the influence of various medium components of the complex formulation; their work led to the formulation of the SP-4 medium, which was developed by maximizing the components found to be useful for spiroplasmas in the Jones study and supplementing it with the CMRL medium used for cultures of vertebrate cells. The SP-4 medium proved to be essential for tickborne spiroplasmas and later turned out to be useful for human and animal mycoplasmas, as summarized by Tully. To date, both approaches have been used for various cultivation needs.

One of the basic criteria for propriety of a named species is that it must be cloned three times before characterization. The reason for this requirement is obvious. In the past, workers have been badly misled by mixed cultures, and of course, taxonomic tests done on a mixed culture are useless. On some occasions, as many as three or perhaps even four spiroplasma species have been isolated simultaneously from a single tabanid fly, and isolations of mycoplasmas from vertebrates have been similarly beset. Nevertheless, workers encountering variable surface antigens, who are concerned that the total antigenic capabilities of a species may not be fully represented by a cloned organism, have recently challenged the usefulness of cloning. But no biological type has ever purported to represent all the genetic capabilities of the entire population. Type specimens always contain a truncated assortment of the population's genetic and phenotypic capabilities, and there is no reason why Mollicutes should be treated exceptionally in this regard. Techniques for cloning Mollicutes that form good colonies on agar have been described by Tully. Many spiroplasmas form poor colonies or none at all; for them cloning in liquid medium on microtiter plates is necessary.

In Spiroplasma, Junca et al. proposed a group system of classification in which Roman numeral designations were given to probable homology groups. Organisms with partial relatedness were classified as subgroups. Organisms with intermediate amounts of DNA-DNA homology and moderate serological interactions with neighboring taxa are eligible for designation as subgroups. Because subgroups have less than 70% shared DNA-DNA reassociation values with putative sister taxa, they, like full groups, are eligible for species designation, provided that they are characterized in accordance with standards established by the minimal standards document. In 1995, the ICSB Subcommittee provided criteria for elevation of subgroups to species. Three groups-group I, group VIII, and group XVI-have been divided into subgroups. Eight of these have been elevated to species level. Later, criteria were established for recognition of new groups (putative species). The requirements that were outlined were as follows. (1) The organism should be triply-cloned by standard techniques. (2) The organism should be demonstrated to be absolutely insusceptible to penicillin. (3) The organism should be shown by thin-section electron microscopy to be bounded by a single membrane and to be devoid of a cell wall and periplasmic fibrils. These three tests would show that the candidate organism was a member of the class Mollicutes and that it was not a spirochete. (4) The organism should be shown to belong to the monogeneric family Spiroplasmataceae; this can be accomplished by dark-field microscopy by demonstrating helicity, at least at some stage of the growth cycle. (5) The generic status should be confirmed by demonstrating the organism's ability to utilize glucose, its failure to utilize urea, and its ability or inability to catabolize arginine. (6) The species status should be determined by showing that the organism is serologically distinct from all other Spiroplasma species or groups by the Spiroplasma deformation test and by at least one other serological technique. If moderate antigenic sharing with another Spiroplasma group or species is observed, it would then be necessary to test antisera directed against the candidate group with antigen of the cross-reacting group. Reciprocal sharing of antigens in these tests would indicate that the candidate organism might be an unrecognized subgroup or that it might be a previously unrecognized serovar of the described group. (7) Determination of the G+C content of organismal DNA is optional but is highly recommended if no plans exist for converting the group into a nomenspecies. (8) The organism must be deposited in a recognized national culture collection. Eventually, groups are eligible for species status. Species descriptions in spiroplasmas, as in all Mollicutes, are strictly governed by the minimal standards document published by the ICSB Committee on Taxonomy of Mollicutes .

Currently, 34 groups and 14 subgroups have been designated, and 34 species descriptions have been published. Twenty-two of these species were designated in the last seven years. For all species designations, the minimal standards document was used as a guideline. The rapid pace of work on this genus has created great interest among microbiologists. Characters used to distinguish Spiroplasma groups and species. Classic characters used for discrimination of Spiroplasma groups and species are: (1) genomic properties: G + C content of the DNA; genome size; and, for intersubgroup comparisons, DNA-DNA hybridization analysis; (2) serology, in which antisera and spiroplasmas are tested reciprocally in metabolism inhibition (MI) and deformation (DF) tests; (3) biochemical and biological properties, including the ability to utilize glucose, arginine, and urea; (4) cultural and other biological properties, including morphology, temperature requirements and doubling time, and organism morphology in liquid culture. Polyacrylamide gel electrophoresis of cell proteins provides good resolution but is rather cumbersome and is used infrequently. Of course, with the advent of a 16S rDNA database for the entire order, sequence data are assuming great importance in classification.

DNA-DNA Hybridization

Because the species concept in Procaryotae, and therefore in Mollicutes and Spiroplasmataceae, is stated in terms of DNA-DNA homology, this test may be required to establish the uniqueness of a candidate taxon. Microbial systematists of all taxa have retained classical DNA-DNA hybridization as a tool of last resort in sorting out intrageneric complexes. DNA-DNA hybridizations have been performed in many recent studies, with or without accompanying sequencing results.

Serology

The criteria for designation of groups require that at least two serological tests be performed to demonstrate the uniqueness of the candidate taxon. Three main techniques have been used for this purpose. Growth inhibition was originally a technique of choice and was used for characterization of groups I-XXIII. This technique is based on the diffusion of specific antibody from a sensitivity disk placed on the surface of an agar plate that had been inoculated with the culture being tested. Measuring the diameter of the zone surrounding the disk in which colonies of the organisms fail to grow assesses results. Of course, this test requires that the test microbe form discrete colonies. Because highly motile spiroplasmas do not form good colonies on agar, some difficulties were experienced with this technique. Although growth inhibition is still used for members of the family Entomoplasmataceae, it was replaced in spiroplasma characterizations by the growth inhibition and deformation tests. D. L. Williamson, who had employed it in early studies with the sex-ratio organism, brought the deformation test into spiroplasmology. The test depends on the ability of specific antiserum to cause abnormal surface structures (blebs) to appear on treated spiroplasmas. In typical tests, the spiroplasmas to be tested are mixed with antiserum at a range of dilutions. After 30 minutes, the reaction mixture is examined by dark-field microscopy. At some ratio of antigen/antibody, about half of the organisms are affected by antiserum. The reciprocal of the antiserum dilution in this mixture is then given as the titer of the reaction. This titer is a clear indicator of the identity of the organism. Homologous reactions usually have endpoints at 1,280 to 20,240, although a few may be as low as 320-640. Strains that are partially related react at titers lower than the homologous titer. The Spiroplasma deformation test is the simplest of the three available serological tests and is therefore the one most commonly employed. However, there are a number of pitfalls in the test, such as the existence of one-way reactions, in which fairly high titers (160-1280) may be observed in one but not both of the antigen/antibody reactions required to demonstrate reciprocal cross-reactivity. Another pitfall is the imperfect morphology of many Spiroplasma isolates during early passage, or even after cloning. In these cases, the organisms may be malformed at the start of the test, and it may be difficult to assess the true effect of antiserum on the organism. The metabolism inhibition test measures the ability of antiserum to inhibit the metabolism (growth in most cases) of spiroplasmas. The diluted antisera are aliquoted into wells of microtiter plates, and the antigen is then added in equal volume. The plates are observed until spiroplasmas in the wells have had an opportunity to grow and cause indicator change in the medium. Positive results for a particular serum are scored in terms of its ability to block the appearance of indicator change in wells. Although very useful, serology is not without its problems. At first, serological methods worked very well with spiroplasmas. Serological methods had been tested against many Mycoplasma species and had proved to be very useful for discriminating among species. And with most spiroplasmas, this proved to be true also. Eventually, however, some cracks appeared. In rare cases, serological tests failed to identify the true group relationship of emerging strains. The first case was that of strain DF-1, which was placed in a new group (XVII) in the 1987 Spiroplasma revision. Eventually, this strain proved to be distantly related to group VIII strains; "bridge" strains appeared. Later, a further member of this group (BARC 2649) was isolated, and again, on the basis of its failure to cross-react with any known member of group VIII, it seemed to be a new group. In the end, however, we became suspicious, given its host origin, morphology, and aggressive catabolism of arginine. Further serological testing uncovered some bridge strains-actually new isolates from tabanid flies in Georgia-that tied B2649 to group VIII. This organism appeared to be unlike any other mollicute identified from plants, and rightly so. In this case, the true identity of the organism was revealed when its 16S rDNA sequence proved to be identical to that of M. iowae. In all these cases, the misleading serological result could have been avoided if a full representation of group antigenicity had been present in the original typing sera. Sometimes, it is used polyvalent typing sera that may contain a mixture of as many as five strains of a given spiroplasma group. Although adequate measures appear to be available for preventing identification failure due to weak serological links, the reverse error-false positive reactions-may be more difficult to deal with. In tabanid spiroplasmas, one group of species seems to have a great propensity to exhibit one-way reactions, in which an organism being tested may exhibit high one-way reactivity with antisera to several different groups. For example, the B4689 strain, when used as antigen, shows reactions of 0.25 to 0.06 of the homologous titer with five heterologous group representatives, all from tabanids. This means, in effect, that when an antigen being tested shows a partial reaction against an existing serum, one can never be sure that the reaction would have been reciprocal had the test been carried out in the reverse direction. Some antigens (group XIX strain PUP-1, group XXVI strain PLHS-1, and, to a lesser extent, group XXVIII strain PALS-1 and group XX strain LD-1), show a proclivity to react with a wide spectrum of antisera. Strain PLHS-1 reacts with almost all *Spiroplasma antisera* that it is tested against. In such circumstances, it is easy to disregard the heterologous reactions, all of which turn out to be one-way when reciprocal tests are performed. In this case, the broad pattern of cross-reactivity was observed in both metabolism inhibition and deformation tests. With the tabanid spiroplasmas, the reaction is much more difficult to interpret. In serotypings of new isolates from tabanids, serological cross-reactivity may be attributable to the existence of discrete serovars of that group or to a propensity to be involved in one-way cross-reactions. If the reaction is one-way, there is no way, given current methodologies, to determine serologically the meaning of the reaction. And there is no way to ascertain if it is a one-way reaction, unless one wishes to prepare antiserum to the strain of unknown identity, a logistical impossibility in most cases.

Transmissible Spongiform Encephalopathy (TSE)

Transmissible Spongiform Encephalopathy (TSE) is identified by the plaques of mutated amyloid protein that form within the brain tissue and destroy synapses and neurotransmitter functions and take on a characteristic sponge or Swiss cheese appearance. Creutzfeldt-Jakob Disease, Scrapie and Kuru are all members of this degenerative disease family, afflictions known about for over two hundred years but not studied intently until the early sixties when they were found to be transmissible.

Dr. Carleton Gajdusek was a young researcher looking for unusual diseases when he visited the Fore Peoples of Papau, New Guinea during the late 1950s. The Papuans of those years were suffering from a population density that put a strain on very limited resources. They practiced female infanticide and cannibalism and were in a constant state of warfare with their neighbors over land and pigs. Severe limitations on normal heterosexual relations were imposed; the men spent most of their time at the men's clubhouse where they prepared for war and engaged in homosexual relations with the young boys. This homosexual activity was all part of an elaborate bonding thought needed to ensure macho warriors and dependable compatriots. The warfare was brutal, often hand-to-hand; capture meant being tortured and killed. Solidarity was essential and achieved through the sharing of semen. The females of the group were disrespected and often abused because they were thought to steal the men's strength and resolve in battle as well as their vital semen. The malnourishment of females and young children was part of this intense process; they supplemented their diet by eating anything that "crawled or crept". Midwives ate placentas of the newborn and women dug up the partly decomposed bodies of relatives and ate and shared with young children the flesh, brains and the accumulations of maggots and mites. This was not a religious ceremony but an attempt to fend off malnutrition.

Gajdusek observed that some of the Fore women and a few children died from symptoms indicating a neurological disorder: dementia, frenzied behavior, blindness and eventual agonizing death. He studied the tribal dynamics and soon hypothesized that the condition, known as Kuru, came from their habit of eating the brains of dead relatives; he brought some diseased brain tissue back to the USA. Gajdusek soon discovered that when he made a broth from the Kuru tissue and injected this mixture into lab animals, they too exhibited the Kuru symptoms. He then processed Kuru diseased lab animals' brains and injected the mixture into other lab animals. They also died the same excruciating deaths. This meant that the condition could be transmitted from organism to organism and was therefore transmissible, hence Transmissible Spongiform Encephalopathy (TSE).

Gajdusek and his colleagues at the National Institute of Health were never able to isolate or positivly identify the agent that causes the TSE even though they've been trying since the early sixties. Scrapie and Creutzfeldt-Jakob Disease were also studied and found to be transmissible. All this was well known underground medical information; many doctors refused to autopsy Creutzfeldt-Jakob Disease victims. For years the NIH conjectured that the infective culprit was a "slow virus". Nothing seemed to distroy the agent; not heat, cold, or any of the normal chemical disinfectants. Nor could they find a trace of its chemical or molecular identity. Furthermore, the virus didn't cause inflamation so antibodies failed to leave a calling card. Some completely new agent was essential.

Another tenacious TSE researcher is Dr. Frank O. Bastian, MD, a professor of pathology and director of neuropathology at the University of South Alabama, Mobile. He has published numerous research articles relating to the etiology of Creutzfeldt-Jakob Disease and also edited a book entitled Creutzfeldt-Jakob Disease and Other Transmissible Spongiform Encephalopathies.

In 1976, Bastian examined a brain biopsy from a patient with Creutzfeldt-Jakob Disease using electron microscopy. He saw a spiral structure foreign to the tissue. It had features of the newly reported spiroplasmas (spiroplasmas were only discovered in 1976). In 1981, a team in New York reported finding a fibril protein in scrapie-infected brain tissue. This scrapie-associated fibril (SAF) protein was 4 nm in diameter and 200 nm long. In 1983, the team looked at various tissues of Creutzfeldt-Jakob Disease and Kuru and demonstrated scrapie-associated fibrils consistently in these diseases but not in control tissues. These SAF were identical morphologically to the internal fibrils of spiroplasmas.

Moreover, antibodies to SAF react with internal fibrillar proteins from Spiroplasma and digested brain material from people with Creutzfeldt-Jakob Disease, suggesting that these proteins essentially are the same. This similarity solidified in Bastian's mind the link between spiroplasmas and Creutzfeldt-Jakob Disease.

Dr. Bastian has postulated that Spiroplasma bacteria causes Creutzfeldt-Jakob Disease and other TSE. His twenty years of research indicates a role for Spiroplasma. The evidence includes the following: spiroplasma-like inclusions were seen in brain biopsies from patients with Creutzfeldt-Jakob Disease (Arch Pathol Lab Med. 1979;103:665-669); spiroplasma internal fibril proteins are identical morphologically to those seen in TSE's; the spiroplasma proteins show immunological cross reactivity with the TSE proteins (J Clin Biol. 1987;25:2430-2431); and spiroplasma, when inoculated into rodents, produces a similar neuropathology. Bastian was able to show that spiroplasmas were neurotropic. When inoculated peripherally into suckling rats, they will eventually localize to the brain tissues. The organisms will produce a persistent infection and produce a spongy change in the brain tissue of these animals. The neuropathologic changes are similar to those seen in Creutzfeldt-Jakob Disease. Spiroplasmas are also within the size range of the agent that transmits Creutzfeldt-Jakob Disease and other transmissible Spongiform encephalopathies. Spiroplasmas will pass through a 50 nm-pore filter. The transmissible agent's size has been determined to be 42 nm.

Prions (PrP)

Bastian also inoculated suckling rats with spiroplasmas, and examined their brain tissues by electron microscopy early in the infection process; he documented the organisms in the tissues. They appeared as membrane-bound forms, except for the one instance where he observed the spiral form. Later in infection, when he knew that the tissues were infectious by broth culture, he couldn't find any evidence of spiroplasmas by looking at the tissues extensively with electron microscopy. Bastian insists that the infection-related protein that most researchers refer to as a "prion" is produced by the host in response to the infection and is not the causative agent. Prions are thought to be self-replicating proteins. Some researchers believe prions are the cause of Creutzfeldt-Jakob Disease and related illnesses because they have found prions in brain tissue from people with Creutzfeldt-Jakob Disease and sheep with scrapie but not in normal brain tissue. Bastian states that a shortcoming in the prion theory is that Creutzfeldt-Jakob Disease and scrapie can be transmitted without prions

Brain material from which the prion has been removed with antibodies can still infect animals. Moreover, the prion has been found in unrelated disease processes, such as Kawsaski syndrome and inclusion body myositis. Prion researchers have jumped to conclusions and have not considered any other possibility. It is quite possible that spiroplasmas may be inducing the formation of the prion protein to protect itself from the immune system. The immune system is very important in the pathogenesis of Creutzfeldt-Jakob Disease. The agent replicates in the spleen and lymph nodes and occasionally causes an immunologic reaction. Auto-antibodies are characteristically seen in the late stages of experimental and naturally occurring disease. The gene for the host protein is located on the chromosome in the region of the major histocompatibility complex (MHC) in the mouse. "The immune reaction seen in these Spongiform diseases can be explained by super antigen activity, Bastian says. He notes that, normally, an antigen is presented to the cell surface in the MHC and interacts with the T-cell receptor-the antigen lying in a groove in the T-cell-MHC sets in motion the standard reaction. A super antigen, on the other hand, binds outside the groove of the T-cell and interacts with the MHC. This results in some immunoglobulin production, but only transiently. The major effect is clonal deletion of T cells, resulting in a state of immune tolerance. Autoantibodies can also form. In Spongiform diseases, PrP presumably acts as a super antigen. It is noteworthy that inclusion body myositis, a condition in which prions are seen, is an established super antigen disease."

Dr. Bastian also notes that investigators have reported transmitting a TSE to mice from hay mites gathered from farms in Iceland where scrapie is endemic (Lancet. 1996). He is virtually certain that these hay mites contain Spiroplasma, noting that the investigators have not so far found PrP in the mites.

6

Plant Diseases

There are several important plant diseases, which are thought to be due to mollecutes. Some of the most common, widespread and economically important diseases are described in this chapter. Diseases caused by non-helicle, non-cultivated MLO include: Alfalfa witches' broom, Banana bunchy top, Clover phyllody, Sesamum phyllody, Sunflower phyllody, Coconut lethal yellowing, Eggplant little leaf, Legume little leaf, Elm yellows or phloem necrosis. Groundnut witches, broom, little peach, Mulberry dwarf, papaya bunchy top,peach decline, X diseases of peach, Potato purple top roll, Potato marginal flavescence, Potato witches' broom, Sandal spike, Sugarcane grassy shoot, Sugarcane white leaf, Tomato big Bud, Tomato purple top, Yellow leaf of areca palm and Yellow wilt of sugarbeet. The diseases are characterized by yellowing, chlorosis, or bronzing of foliage, stunting (shortening of internodes and reduced size of leaves), proliferation of axillary buds resulting in witches' brrom effect, proliferation of secondary roots and abnormal fruits and seeds. Flowers are often sterile. Some diseases show few of these symptoms whereas others may show all or most of these symptoms. Variation in symptom syndrome may occur according to host species, the time and mode of infection, environmental factors, and mixed infection by more than one type of phloem-limited pathogens such as dual infection of MLO and Spiroplasma or MLO and RLB or other bacteria or even virus. Established spiroplasma diseases also show most of these symptoms but virescence is not caused by spiroplasmas. The disease caused by *Spiroplasma* spp. Include

Bermuda grass white leaf, Citrus stubborn (*S. citri*), corn stunt (*S. kunkelii*), Opuntia witches' broom, Rice yellow dwarf, and yellows of aster and lettuce.

Aster Yellows Disease

Aster yellows is a disease that affects over 300 species of plants, including flowers, vegetables, and weeds. Susceptible flowers include aster, chrysanthemum, cockscomb, coreopsis, cosmos, daisy, dianthus, echinacea (coneflower), gladiolus, marigold, petunia, and phlox. Susceptible vegetables include carrots, onions, potatoes, and tomatoes. Weeds such as dandelions, plantain, and thistle are also susceptible and can serve as a source of inoculum in home gardens.

Causal Organism : *Phytoplasma*

Vector

Leafhoppers. The six-spotted leafhopper is the main vector, which transmits the *Phytoplasma* that causes aster yellows in flax and in many other crops such as canola and sunflower, and in some weeds.

Disease Cycle

Aster yellows is primarily transmitted by insects called leafhoppers, and through grafting. The phytoplasma survives winter in perennial and biennial plants. In the spring, leafhoppers feed on infected plants. During feeding, plant sap containing the phytoplasma is sucked into the leafhopper's body where multiplication of the microorganism takes place. Following an incubation period, the phytoplasma is transferred to healthy plants during leafhopper feeding. When a leafhopper feeds on a plant infected with aster yellows it becomes "infected" with the phytoplasma and remains infected throughout its life. The phytoplasma cells multiply and cause infection of the insect's salivary glands within one to three weeks. When the infected insects feed on healthy plants, they inject the phytoplasma cells into the plant phloem. Susceptible plants will be symptomatic in 10 to 40 days.

Fig. 6.1. Aster yellows in field

Fig. 6.2. Aster yellows; flowers are sterile.

The spread of aster yellows is worse in cool wet summers. Hot dry weather is not favorable for either the phytoplasma or the leafhopper. As with many disease and pest problems, diagnosis is perhaps the most important factor in controlling aster yellows. The cause of aster yellows is a microscopic organism called a phytoplasma (formerly called a mycoplasma-like organism or MLO). Phytoplasmas are most similar in size and composition to bacteria.

Symptoms

Specific symptoms vary from plant to plant. Symptoms of aster yellows are more severe and appear more quickly during warm weather. At lower temperatures, plants may be infected without symptom expression. There are following Aster yellows symptoms:

1. The roots are bitter, stunted, and deformed with tiny, hair-like roots developing all over the main root.
2. Infected plants may also produce a proliferation of spindly, secondary shoots, giving the plant a bushy appearance.
3. Yellowing of the top part of the plant, yellowing of the foliage, Carrots and marigolds can be severely affected by aster yellows. Leaves of infected carrots grow in tight bunches. The inner leaves are yellow and stunted, while outer leaves turn rusty red to reddish purple. Infected marigolds have small, yellow leaves.
4. Chlorosis, yellowing of the leaves while the veins remain green, is a major symptom of aster yellows. Growth slows down and leaves may be smaller and more narrow than usual. Foliage is sometimes curled.
5. The symptoms of the disease will often differ depending upon what species is infected. For instance, carrot roots may be bitter and hairy while lettuce may show pink or tan spots and have twisted inner leaves.
6. Production of many spindly stems and/or flower stalks.
7. Conspicuous malformation of the flowers, all flower parts including the petals is converted into small, yellowish green leaves.

8. Diseased flowers are sterile and produce no seed. Flowers often fail to develop color, remain green and distorted, and seeds or fruit never develop. Flowers that are produced are often irregular, deformed, and green. Flowers may be deformed and exhibit bizarre tufts of deformed leaves inside the flower or in place of the flower.
9. The severity of the disease depends on the stage at which plants become infected and the number of insect vectors that carry the organism. The whole plant is generally stunted and spindly.

Control measures

1. Once a plant is infected with aster yellows, it is a lost cause since the disease is incurable. Early diagnosis and prompt removal of infected plants may help reduce the spread of the disease. Although the disease itself is not fatal to the plant, its presence makes it impossible for a plant to fulfill its intended role in the garden.
2. Controlling aster yellows is difficult. As long as infected leafhoppers are around, they can infect plants. A practical way to avoid having problems with this disease is to grow plants that are not as susceptible to aster yellows. Verbena, salvia, nicotiana, geranium, cockscomb, and impatiens are among the least susceptible plants.
3. Vegetable growers may protect susceptible crops by using the mesh fabrics that keep leafhoppers and other insects away from the plants. Some growers put strips of aluminum foil between rows because bright reflections of sunlight confuse the leafhoppers.
4. Remove weeds in lawn, garden, and surrounding areas, including plantain and dandelion that may harbor the disease. Insecticides are generally not recommended for the control of leafhoppers in the home garden.
5. Seed early to avoid the migrating leafhoppers in mid and late season. Seeding early also reduces the incidence and severity of aster yellows.

Papaya Bunchy Top Disease

Papaya bunchy top has been observed throughout much of the Caribbean region, from Cuba southward to Trinidad, since it was first reported in Puerto Rico in 1931. Bunchy top is a devastating disease and can severely limit commercial papaya production.

Causal Organism : *Phytoplasma* (MLO)

Vector

Leafhoppers. Two leafhoppers, *Empoasca papayae* Oman and *E. stevensi* Young, transmit the bunchy top agent, and the occurrence of the disease coincides largely with that of these vectors. *Empoasca papayae* has most frequently been associated with the disease. *Empoasca papayae* has rarely been found on other plant species and is the only leafhopper known to breed on papaya. *Empoasca stevensi* was originally described from specimens collected from papaya in Florida in 1940, but little is known about the ecology and distribution of this species.

Disease cycle

Single insects of both vector species can transmit the bunchy top MLO. The MLO also can be transmitted by grafting. Symptoms usually begin to appear 30 to 45 days after inoculation. Some papaya cultivars are more tolerant to the disease than others, but immunity is not known. The degree of tolerance affects the rate and severity of symptom expression. However, it is not known whether vector preference and the ability to transmit the bunchy top MLO are affected by cultivar differences. Although bunchy top was originally thought to be caused by a virus, the disease is most likely caused by a mycoplasma-like organism (MLO). MLOs have been observed in the phloem of infected plants using transmission electron microscopy, and infected plants treated with tetracycline-based antibiotics exhibited a remission of symptoms. Although the absence of latex flow from fresh puncture wounds in leaves, petioles, stems, and fruits were considered diagnostic for bunchy top, a more recent study indicated that this technique is unreliable. The presence of MLOs in plants with foliar symptoms of bunchy top and normal latex flow was confirmed by electron microscopy. The DNA-binding fluorochrome, 4,6-diamidino-2-phenylindole (DAPI), can be used to stain MLOs in the phloem of numerous plant species, thus, making their presence detectable by epifluorescent microscopy. This technique has been used to detect MLO's in petiole and leaf vein tissue of papaya trees with bunchy top symptoms. The reliability of this method to diagnose bunchy top has not been established.

Fig. 6.3. Papaya bunchy top.

Symptoms

1. The first symptom of bunchy top is a faint mottling of the upper leaves.
2. The laminae of infected leaves progressively become more chlorotic, especially in interveinal areas, and may eventually exhibit marginal necrosis.
3. Leaves and petioles show reduced growth, and become rigid.
4. Internodes become progressively shortened, and petioles tend to assume a more horizontal position.
5. Oily appearing spots are often present in the upper parts of stems and in petioles.
6. Apical growth ultimately ceases, which, with the shortening of internodes, imparts a "bunchy top" appearance to affected plants.
7. With the cessation of apical growth and decline of plant vigor, the oldest leaves progressively abscise, leaving fewer, more stunted leaves at the top. Eventually, plants may be entirely killed; however, new shoots may sprout from the lower, non-affected portion of stems.

8. If fruits are set on infected plants, their flavor may be bitter due to the disease.

Control measures

1. Currently, the use of tolerant cultivars is the only practical means of control of bunchy top, but it has limited application. Where bunchy top occurs, tolerant cultivars can only be grown commercially in locations with low disease pressure. Disease pressure varies with geographic location, presumably due to insect vector ecology.
2. Application of insecticide to reduce the incidence of leafhopper vectors may be beneficial.
3. In the past, the disease has been managed through a program involving leafhopper control throughout the productive life of the crop and the removal of sources of inoculum. Rouging infected trees or the topping of infected plants below the point where latex exudes after wounding attained the latter. Axillary shoots that develop after topping are often free of the disease. However, the success of this program depended largely on the availability of a pesticide with long residual activity, namely DDT, which is no longer approved for use in many areas. A suitable replacement for DDT has not been identified.
4. Antibiotic therapy is an effective control measure from an experimental standpoint but has not been applied commercially.
5. Drenching the soil around infected plants with chlortetracycline or tetracycline hydrochloride and drenching combined with root-dip treatments were both successful. The efficacy of foliar application or trunk injection has not been established.

X-Disease in Peaches

X-disease has been diagnosed in several peach orchards in worldwide. Once the disease is established in an area or orchard, it can be very destructive.

Causal Organism : X-disease mycoplasma

Vector : Leafhoppers.

Disease cycle

Peach X-disease often follows a 10-15- year cycle. The disease will be very severe for 4-6 years, then will gradually disappear as a commercial problem until the next cycle begins. A good diagnostic symptom is the presence of apparently healthy twigs or branches with normal-looking leaves mixed with twigs or branches showing the symptoms described earlier. This mixture of healthy and diseased branches on the same tree occurs primarily during the first and second years of infection. Two to three years after initial infection, most branches will show symptoms. X-disease is caused by a phloem-limited mycoplasma - a minute pathogenic organism smaller than most bacteria. The X-disease mycoplasma is transmitted by at least eight species of leafhoppers. White apple leafhopper, rose leafhopper, and potato leafhoppers are not vectors. In fact, none of the X-disease vectors are abundant enough to cause direct feeding damage and they usually escape notice in the orchard. However, they are very efficient vectors of X-disease. The leafhoppers acquire the X-disease organism while feeding on diseased chokecherry bushes, on infected sweet cherry trees, or on wild seedlings of sweet cherry. They do not acquire the X-disease mycoplasma from diseased peaches because the population of the disease organism within diseased peach trees is so low that leafhoppers do not encounter the organisms while feeding. X-disease can be transmitted by several species of leafhoppers. These leafhoppers acquire the X-disease pathogen while sucking juices from the leaves of X-disease-infected choke-cherries. Wild chokecherry is an important reservoir for the X-disease mycoplasma. Two to three weeks later the leafhopper can inject the agent into healthy leaves while feeding. These leafhoppers are usually not considered pests of peach and cherry. The main damage they cause is the incidental transmission of X-disease. Movement of leafhoppers, and therefore X-disease, can also occur from infected sweet and sour cherry, particularly from trees. Although the possibility of spread from peach to peach has been investigated many times, it appears to be of minor importance. After leafhopper vectors feed on an infected plant, the X-disease organism must grow within the insect for at least 20 days before the insect can transmit X-disease to another plant. Once that 25-day incubation period is completed, however, the leafhoppers with X-disease remain infective for the rest of their lives. In laboratory studies, leafhoppers have often lived for 30-40 days after they become infective. A single infective insect therefore has the potential to infect numerous plants. Peach trees that become infected with X-disease are usually inoculated by leafhoppers during July and August, and symptoms then develop on the trees the following year. Leafhoppers can continue to transmit X-disease to peach trees during September and early October, but many of the late-season transmissions fail to cause disease because the pathogen does not become established in the plant following late-season transmissions. Relatively mild winters during the last two years may have allowed more of the late-season transmissions to persist through winter and may therefore have contributed to the increasing severity of X-disease.

Fig. 6.4. Symptoms of X-disease on peach leaves.

Symptoms

Symptoms on peach may be variable, depending upon the environment, variety, and possibly other factors. However, once acquainted with general symptoms, one should be able to recognize the disease more often than not.

1. Symptoms are predominantly foliar, but the fruits may also be affected. The first symptoms occur on foliage. X-disease causes leaves on infected peach trees to turn yellow, curl upward, and develop red, water-soaked spots that are not limited by leaf veins. The leaf disorder results in early defoliation of the oldest leaves, leaving a "horse-tail" of young foliage at the end of affected terminal shoots.
2. Leaves on affected branches fall prematurely, starting at the base of the branch. Eventually, only a tuft of leaves remains at the tips of infected shoots.
3. Fruit set on trees infected with X-disease may appear normal at first, but fruit on infected branches will usually drop prematurely.

Control measures

1. Reducing populations of leafhopper vectors is an approach to X-disease control. Maintaining a vigorous insect control program from June through harvest with insecticides effective against leafhoppers should be useful in reducing spread of the disease.
2. Remove infected trees near peach orchards.

Leaf Variegation or June Yellows Disease in Strawberries

Although leaf variegation or June yellows in strawberries was noted and described as "gold-striped" and "silver-striped" as early as 1719, the cause of the disorder is still not known. Various synonyms for the disorder are Blakemore yellows, spring yellows, chlorosis, and noninfectious variegation. Leaf variegation occurs commonly in, but is not limited to, strawberry cultivars that have Blakemore, Howard 17 (Premier) or Auchincruive Climax in their parentage.

Causal organism : Mycoplasma-like organism (MLO)

Vector : Grafting or sap inoculation.

Disease cycle

No causal agent has been associated with leaf variegation or June yellows. The disorder has not been transmitted asexually by grafting or sap inoculation but can be transmitted sexually to seedlings of affected parents. Its appearance is unpredictable in all breeding lines and may not become evident in a particular clone for several years. The current theory is that leaf variegation is due to some entity within the strawberry host cells that acts similarly to certain virus or mycoplasma infections. Electron microscopy has thus far not identified a virus or mycoplasma in diseased plants.

Fig. 6.5. Leaf variegation or June-yellows; healthy leaf to right.

Symptoms

1. Symptoms first appear in the spring on new unfolding leaflets. A puckering and distortion may occur in leaves with a "white streak" type of variegation in early stages of the disease.

2. The leaves become irregularly mottled, streaked, or spotted with golden or pale yellow-to-white and light green areas. A slight loss of plant vigor accompanies the color change.
3. Variegation occurs mainly during cool weather, when the temperatures are below 50°F (10°C), in the spring or fall, although some cultivars may remain mottled throughout the summer.
4. Variegation has been observed only on the foliage and flowers - never on fruit. All runners and daughter plants produced by diseased plants are also variegated with no reversion to a normal green color.
5. Seemingly healthy green plants may become variegated at any stage. There is no way to predict when a plant or its offspring will become variegated.
6. As the disease progresses, the symptoms increase in intensity. The leaves become progressively more mottled each year until they are completely golden yellow-to-white and frequently puckered or otherwise distorted.
7. White-streaked or fully yellowed leaves never regain a normal green color during the summer.
8. Affected plants become dwarfed.
9. Fruit production is reduced considerably, with the fruit being small and of poor quality.
10. Affected plants never recover and usually die within two to three years.
11. Permanent white streaks or sectors, with or without accompanying mottling symptoms, may occur in some cultivars affected by leaf variegation or June yellows. White streak may be a chimera (the result of a somatic mutation). It occurs more commonly in some cultivars, for example, in Earliglow, than in others.

Control measures

1. Do not propagate from variegated mother plants.
2. The only control to save a variety from extinction is to locate a source of "yellows-resistant" or normal plants of the desired variety. These plants, if found, should be propagated by virus-free procedures to increase the number of healthy plants. Never accept plants that show leaf variegation.

3. The disorder appears to be noninfectious in that normal green clones may grow beside variegated clones for long periods without becoming variegated.
4. Where the percentage of variegated plants in new or established plantings is reasonably low, variegated plants should be rogued (removed and destroyed) as soon as the disease is detected and replaced with nonvariegated, certified, virus-free plants.
5. At present there is no cure for leaf variegation, and there is no assurance that healthy green plants in a variegation-susceptible cultivar will remain nonvariegated.

Corn Stunt

Causal Organism : Mycoplasma, *Spiroplasma kunkelii*

Vector : Corn leafhopper (*Dalbulus maidis*)

The adult corn stunt leafhopper and nymphs are light colored. Adults fly or jump when disturbed and often gather on the corn plant in the center of the whorl. Nymphs have no wings and readily move when disturbed. Early planted corn is rarely affected. Leafhoppers suck juices out of leaves with their stylet mouthparts. High populations can cause premature leaf drying, and black sooty mold may grow on the honeydew. The presence of leafhoppers however does not necessarily mean corn stunt occur. The insect only develops high populations in some years and vectors the corn stunt pathogen even less often.

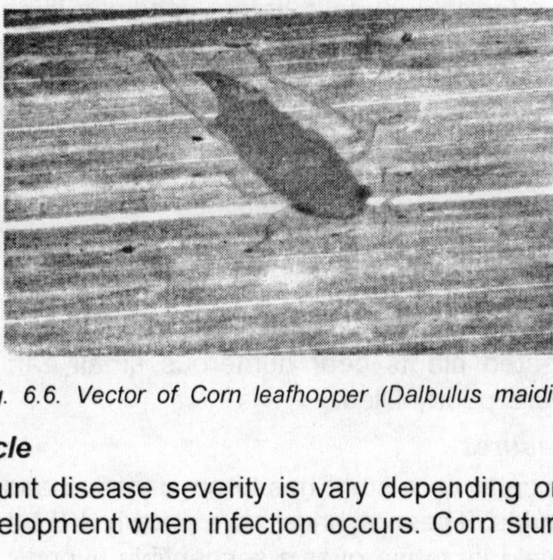

Fig. 6.6. Vector of Corn leafhopper (Dalbulus maidis).

Disease cycle

Corn stunt disease severity is vary depending on the stage of plant development when infection occurs. Corn stunt is caused

Fig. 6.7. Corn stunt disease.

by a mycoplasma, *Spiroplasma kunkelii*. Mycoplasmas differ phenotypically from other bacteria by their minute size and lack of a cell wall. The disease is transmitted to a plant when the vector, the corn leafhopper (*Dalbulus maidis*), feeds from the plant phloem. The corn stunt mycoplasma is introduced into a healthy plant with the insect saliva. In the plant, the mycoplasma is restricted to the phloem, but spreads throughout the plant infecting all areas. Corn leafhoppers carry the disease from infected corn to healthy corn. *S. kunkelii* is not transmitted via plant seeds, nor are the leafhopper eggs infected. The leafhoppers feed on an infected plant for several days before acquiring the spiroplasma. It then requires a two to three-week incubation period before the leafhopper can infect a healthy plant. It cannot survive outside its host and is dependent upon transmission for survival and dispersal.

Symptoms

1. The infected plants are stunted.
2. Leaves have broad chlorotic strips and with age they take on a reddish-purple colour.
3. Internodes are reduced in length.
4. The infected plants bear numerous small ear shoots that usually are poorly filled.

Control measures

Resistance to corn stunt has been identified and is available in some hybrids. These hybrids have shown a 0.5 to 0.75 tons/acre grain yield increase over a susceptible hybrid.

Tomato Yellows Disease

In some years, serious outbreaks of tomato yellows have occurred in southern British Columbia. The severity varies from year to year; with a high infection 30%, or more, of the crop may be lost. The disease is widespread throughout the Pacific coast region extending from British Columbia to Mexico.

Causal organism : Beet curly-top mycoplasma.

Vector : Beet leafhopper.

Symptoms

1. Diseased plants are readily distinguishable in the field, because they are rigid and have stiff brittle leaves that tend to roll inward along the midrib.
2. They curve downward slightly and have a drooping but not wilting appearance. The veins are generally, but not invariably, purplish and the normally green leaves change to sulfur yellow.
3. The infected plant ceases to grow and the fruit already formed does not enlarge.
4. The pith dries out and the stem becomes hollow. Eventually the plant turns brown and dies.
5. The disease runs its course in 1-2 weeks. Affected plants may recover somewhat but never regain a normal appearance. The small roots become brown and decayed.

Control measures

This disease is caused by the same mycoplasma that causes curly-top of beets; therefore always separate crops of tomatoes from beets. Maintain good weed control.

Purple-top Disease of Tomato

Causal organism : Aster yellows mycoplasma

Symptoms

1. The most characteristic symptom is the purplish tinge in the foliage, which is particularly apparent in the youngest leaves.
2. Affected plants show an upright appearance because of slight hardening of the stem and leafstalks.
3. The leaves also show a tendency to roll slightly.
4. Considerable stunting of growth occurs.

Control measures

Purple-top disease is rarely serious enough to warrant specific control measures. Good weed control is always sound husbandry; it removes sources of inoculum.

Yellows Diseases

Plant pathogenic microorganisms called phytoplasmas are the major cause of yellows diseases in plants. "Yellows" diseases are the most important group of diseases spread by phytoplasmas. Phytoplasma causing the yellows diseases of echinacea, monarda and caraway described.

Causal organism : Phytoplasmas

Vector : Phloem-feeding leafhoppers.

Disease cycle

Phytoplasmas are transmitted between plants primarily by phloem-feeding leafhoppers (Insecta: Homoptera: Cicadellidae). These insects have a well-developed sensory system that enables them to reject unsuitable host plants without sampling the sap. They usually transmit phytoplasmas most efficiently to plants preferred for feeding. Leafhoppers feed on a variety of plants and can spread phytoplasma pathogens from wild to cultivated plants and vice versa. Infected biennial and perennial plants may allow phytoplasmas to persist from one year to another in fields and natural areas.

Fig. 6.8. A leafhopper can transmit phytoplasmas efficiently to the plants it feeds on.

Leafhoppers can survive through the winter as eggs and appear as early as May. The incidence of phytoplasma diseases often depends upon the abundance and time of migration of leafhoppers from infested to healthy plants. Traditional field crops, such as alfalfa and potato, may serve as symptomless carriers of phytoplasma pathogens and also harbour adult leafhoppers, which could transmit primary inoculum to adjacent plantings of specialty crops.

Fig. 6.9. Phytoplasma-infected monarda plant.

Fig. 6.10. A phytoplasma-infected echinacea plant with buncy, red leaves.

Symptoms

1. Affected plants generally show leaf yellowing, reddening and stunting in the early stages of disease development.
2. The yellowing symptom can be quite striking, and the name "yellows" is often used to describe these diseased plants.
3. Plants with yellows sometimes have extremely numerous, small, branched, axillary shoots coming from the stem nodes, giving them a bunchy or witches'-broom appearance. In later stages of symptom development, infected flowers may change into vegetative, leaflike structures (virescence and phyllody) and, consequently, fail to produce seeds.
4. Yellows can also reduce winter hardiness, and it often kills affected plants.

Fig. 6.11. Phytoplasma-infected echinacea plants showing abnormal flower development ray and disk flowers of a phytoplasma-infected echinacea plant have turned green and leaflike.

Fig. 6.12. Phytoplasma-infected caraway flower heat (left) showing bunching, rosetting and malformation of floral organs (virescence and phyllody).

Control measures

1. An integrated strategy, focussing on both the pathogen and the vector, must be followed to successfully manage phytoplasma diseases in susceptible crops. Susceptible specialty crops should not be planted close to fields of forage legumes or other perennial crops affected by phytoplasma diseases.
2. Care should also be taken to control perennial weeds that may harbour phytoplasmas and/or leafhoppers both in susceptible crops and in adjacent fields and around their margins.
3. Specialty crops and surrounding fields and headlands should be regularly monitored for leafhoppers.

4. Keeping leafhopper populations low is an important means of controlling phytoplasma diseases.
5. Registered insecticides can be used to control leafhoppers in some situations.
6. Development of resistant cultivars is the most promising method for controlling phytoplasma diseases in plants, but little effort has been devoted to breeding specialty crops with resistance to phytoplasmas.

Potato Purple-top Wilt MLO Disease

This disease is particularly Occurs in North America and, in so far as it causes potato haywire disease, is considered as an A1 quarantine pest by EPPO (OEPP/EPPO, 1984). All but the first of these are regarded as belonging to "the aster yellows" group, with yellow or purple leaf discoloration as main symptoms. They are not usually transmissible by tuber. The leafhopper vectors include Macrosteles and Hyalesthes spp. In Europe, there are numerous MLOs of the aster yellows type, attacking various hosts, but only potato stolbur MLO is found in potato. It is usually considered as a distinct MLO. Tomato big bud MLO occurring in Australia, several Asian countries and USA causes a disease very similar to stolbur. The three Indian MLOs are also recognized as of the Old World aster yellows type and are distinguished from potato stolbur MLO. This disease, reported for the first time from India, can cause 15 to 80 percent loss in yield.

Causal organism

Mycoplasma Like Organism (MLO) of the aster yellows complex.

Vector

Leafhopper Macrosteles fascifrons and nymphs of Alebroides nigroscutellatus (*A. dravidanus*)

Disease cycle

Under natural conditions, potato purple-top wilt MLO can be dispersed by its leafhopper vectors. Internationally, it could be carried in potato plants or in vector insects associated with them, but such movement hardly occurs in practice. It should be stressed that aster yellows type MLOs are not normally tuber-transmitted in potato. The principal vector is the leafhopper Macrosteles fascifrons, in which the MLOs are propagative. The vectors remain infective for life. They can feed on a variety of plants, and the apparent host range of the MLO depends more on the host preferences of the vector than on any particular specificity between plant and MLO. Potato purple-top wilt MLO from Japan is reported to be transmitted by Scleroracus flavopictus and not by Macrosteles spp., unlike aster yellows MLO from USA. Potato MLOs vary in their size and shape. They are pleomorphic with a diameter up to 1000 nm. They occur mainly in the sieve tubes and can be seen there by electron microscopy or fluorescent detection methods. In India, Nymphs of Alebroides nigro-scutellatus (A. dravidanus) are better transmitters of the pathogen. The MLO adversely affects its vectors. The urge for mating in the females is delayed and often they become sterile. The hatching time of eggs is enhanced. The longevity of adults is also reduced. The minimum acquisition access period is one hour. Maximum transmission takes place with acquisition access period of three dyas. The MLO needs 7 to 15 days of incubation period in the vector body. One or a few infective vectors are required per plant for maximum transmission. Nymphs are more efficient than adults in transmitting the MLO. While adult males have better transmitting ability than females, the latter retain the MLO in their body for a longer time. Infected tubers seem to be the most important source of perennation and transmission. The symptoms are pronounced at low temperatures of 150 to 200°C and masked at 300 to 380°C. The MLO can be transmitted to clover, tomato, Datura and tobacco.

Symptoms

1. Apical leaves are pinched or curled, roll inward and develop yellowish-purple pigments.
2. Infected plants lose their apical dominance and proliferation of axillary buds occurs.
3. Plants may wilt and die prematurely. These plants commonly form hair sprouts.
4. According to Conners (1967) the primary symptom is purple-top wilt, and the haywire stage is due to secondary infection.
5. It is characterized by rolling and purple or pink colouration of the basal part of the leaflets of top leaves. Some cultivars develop only rolling but no colouration. Stunting, chlorosis, profuse axillary shoots with aerial tubers, and swelling of nodes

are other symptoms. There is no wilting of the infected plants and no necrosis of the tubers (Giri and Nagaich, 1971; Nagaich and Giri, 1973).

Control measures

The diseases are not important enough to merit any particular control measures.

Grassy Shoot Disease of Sugarcane

Grassy shoot, yellowing and white leaf (albinism) diseases of sugarcane in India, Taiwan and Thailand have been considered as the same disease caused by the same organism. Grassy shoot was first noticed in India in 1919 in Maharashtra and is one of the serious diseases of sugarcane there. It has been found in Andhra Pradesh, Tamil Nadu, Orissa, Bihar, Uttar Pradesh, Uttaranchal, Punjab and Rajasthan. Outside India, Grassy shoot is reported from Burma, Sri Lanka and Sudan also.

Causal organism : Mycoplasma Like Organism (MLO)

Vector

Rhopalosiphum maidis (*Aphis maidis*), *Aphis sacchari*, and *Aphis indosacchari* and leafhopper.

Disease cycle

The organism is present in the sieve tubes of the phloem as ovoid, spherical, or irregularly shaped bodies. The size of ovoid or spherical bodies is 300 to 400 nm in diameter and that of protruding filaments 30 to 50 nm in diameter. These cells are devoid of true cell wall and are bounded by a single triple layered membrane. They contain ribosomal granules and DNA strand. Mostly the cells are concentrated towards the periphery of the cell, near the cell-walls but often the cells are fully loaded with these bodies and then the cell dies followed by death of the MLO due to starvation. In some instances MLO are seen lying close to sieve pores of the sieve tubes and turn filamentous while passing through the sieve pores. Concentration of MLO is high in canes growing at high temperature (around 300^{0}C) and such canes show severe symptoms. At lower temperatures the symptoms are less severe and the number of MLO in the cells is also lower.

The white leaf disease of sugarcane reported from Taiwan has some difference from grassy shoot disease described in India. While grassy shoot organism is inactivated by hot water at 500^{0}C for two to three hours the white leaf disease agent is inactivated by hot water at 550°C for 30 minutes or 540°C for 40 minutes. Aphids transmit the grassy shoot while leaf hoper transmits white leaf.

The grassy shoot MLO perpetuates through diseased canes used for seed and spread through diseased sets and cane cutting knives (sap transmission). Ratooning of a diseased crop is an important source of perennation. Many aphids and a leafhopper have been reported to spread the MLO from plant to plant in the field but only to a limited extent. The vectors are *Rhopalosiphum maidis* (*Aphis maidis*), *Aphis sacchari* and *Aphis indosacchari*. The minimum acquisition feed period is 34 seconds and inoculation feed period 30 seconds. However, optimum acquisition feed period is 15 minutes. Although a single infected aphid can transmit the pathogen, optimum number is 15. Transmission is in the non-persistent manner. Winged forms are more effective in spreading the disease in the standing crop. The disease has been transmitted from sugarcane to sorghum and from sorghum to sugarcane by using Aphis sacchari (*Melanaphis sacchari*). Sorghum is also a natural host of the MLO.

Symptoms

The disease shows different types of symptoms at different stages of plant growth and, thus, has been variously described as 'new chlorotic disease', 'albino disease', 'yellowing disease', 'bunchy disease' or 'leaf tuft', Profuse tillering is the main symptom and, therefore, the standard name for the disease is grassy shoot.

1. The most pronounced symptom is the grassy appearance of the affected stools.
2. Shoots growing from diseased setts remain dwarfed or stunted.
3. The leaves are narrow and small, like grass leaves; the canes are thin with short internodes, giving a bunchy or grassy appearance to the clump.
4. The leaves appear yellowish and in some cases may be entirely devoid of any pigment (white leaf or albinism). If many of the tillers are affected in this manner the entire stool dries.
5. In secondary infection, ordinarily the new leaves and internodes show the above symptoms. Leaves exhibit straight, long, white or light green or yellowish streaks.

6. The lower nodes produce large number of grassy shoots. In systemically infected canes the disease appears in May-June. In sprouts raised from top buds the symptoms appear late as compared to sprouts raised from lower buds. This suggests difference in the concentration of the causal agent in the cane.

Control measures

1. Healthy seed sets should be taken from a field where grassy shoot is not present even in traces.
2. During the early stage of crop growth insect vectors can be controlled by weekly spraying of 0.16 per cent malathion.
3. Diseased clumps should be dug out and destroyed.
4. Heat therapy has also been recommended. The sets should be dept in hot water at 500^{0}C for two hours or in hot air at 540^{0}C for eight hours. This treatment is feasible only for seed nurseries.
5. Tetracycline antibiotics (250 ppm) applied to seed sets under negative pressure have been found to eliminate symptoms. Single bud sets can be treated with such antibiotics.

Little leaf of Eggplant (Brinjal)

Little leaf of brinjal (*Solanum melongena*) is found throughout India and Sri Lanka and probably occurs in other neighbouring contries. In India it was first reported from Coimbatore (Tamil Nadu). The disease has become a serious threat to the profitable cultivation of this vegetable crop in most of the states. When young plants are attacked they do not produce flowers and fruits.

Causal organism : Mycoplasma Like Organism (MLO)

Vector : *Cestius phycitis* or *Hishimonus (Cestius) phycitis*

Disease cycle

In nature, the causal agent of brinjal little leaf is transmitted by the vector *Cestius phycitis* or *Hishimonas phycitis*. The same MLO occurs on *Datura fastuosa* and *Vinca rosea*. Artificially it has been transmitted to tomato, potato and tobacco. Probably, during the off-season of the brinjal crop the causal agent survives on weed hosts and from there it is transmitted to the main host by its insect vector.

Symptoms

1. The main symptom of the disease is production of very short leaves by the affected plant.

2. The petioles are so much reduced in size that the leaves appear sticking to the stem.
3. Such leaves are narrow, soft, smooth and yellowish in colour. Newly formed leaves are further reduced in size.
4. The internodes of the stem are shortened and at the same time a large number of axillary buds are stimulated to grow into short branches with small leaves. This gives the whole plant a bushy appearance.
5. Usually such plants fail to form flowers. Even if flowers are formed they remain green.
6. Fruiting is rare.

Control measures

No effective control measures against the disease are known.

1. Atibiotic treatments with tetracyclines have been claimed to control the disease.
2. In the absence of any definite control measure eradication of weed hosts and diseases brinjal plants and control of insect vectors by insecticides are the only recommendations against the disease.

The Sandal Spike Disease

Sandal (*Santalum album* L.) is an angiospermic semi-root parasite of many plant species, which grows to the size of a tree. It is one of the most economically important forest trees in India. The species is confined to India and Indonesia. In India the tree is common in the southern parts including states of Karnataka, Tamil Nadu and Kerala. Spike disease is the most destructive of the few diseases that attack this plant. It is a yellows type of disease with witches' broom effect. Trees of all ages and size are attacked and die within a few years of appearance of sysmptoms. The disease is most common in Karnataka but has also spread to parts of Tamil Nadu and Kerala.

Causal organism : Mycoplasma Like Organism (MLO)

Vector

Dodder, root contacts, insect vectors and leafhoppers *Nephotettix virescence* and *Moonia albimaculata*.

Disease cycle

The spike disease is transmitted through root contacts and insect vectors. More than one species of leafhoppers seem to be involved in its transmission. These include Nephotettix virescence and Moonia albimaculata. The size of MLO in the host ranges from 60 to 750 nm, the most commonly occurring form being ellipsoidal, measuring 180-220 x 250-300 nm. Some cells are elongated, measuring up to 750X150nm. The MLO grows best at 300 to 3800°C. Below and above this range the growth declines. The MLO is reported to infect a large number of plant species such as *Eucalyptus grandis*, *Vinca rosea*, *Zizyphus oenoplea*, *Dodonea viscose*, etc. The disease can be transmitted from vinca rosea to sandal and vice versa through dodder.

Symptoms

Generally two types of symptoms are produced. These are 'rosette spike' and 'pendulous spike'.

1. The more common symptom is 'rosette spike', in which found severe reduction in leaf size and shortening of the internodes. This results in crowding of leaves on the leaf bearing branches.
2. The new leaves that develop later are further reduced in size. The reduction is more in width than in length. Such leaves stand out stiffly on the branches like spikes.
3. In advanced stages of the disease, just before death of the tree, leaves become yellowish and finally reddish.
4. The flowers of the infected trees show phyllody (virescence).
5. The diseased parts rarely bear any fruit although the healthy parts of the same tree bear normal flowers and fruits.
6. Ends of roots of diseased plants die and haustorial connections with the host plant are damaged.
7. Phloem necrosis is one of the internal symptoms.
8. The 'pendulous spike' is another characterized symptom of this disease in which continuous apical growth of individual shoots without proportionate thickening of the shoot resulting in a drooping habit of the shoot.
9. The dormant buds do not develop or grow and, therefore, no resetting effect is seen. In this form of the diseased roots and haustoria are not damaged.

Control measures

So far no specific measures for control of sandal spike disease have been developed.

1. Attempts had been made to cure the trees by raising temperatures above 380^{0}C or by boiling fires in trenches around the affected trees but it was neither successful nor practical.
2. Eradication of diseased trees and creating disease free strips around affected areas have also failed to arrest its spread. No host resistance has been found.
3. Like other plant diseases of MLO etiology sandal spike also shows remission of symptoms after treatment with tetracycline. The systemic fungicide Benlate was also reported to bring about temporary remission of symptoms.
4. Application of the antibiotic by the girdling method shows recovery of the trees 25 to 30 days after treatment with terramycin alone or in combination with Benlate. However, these treatments result in only temporary relief for 50 to 150 days and then the disease reappears.

Citrus Greening Disease

Citrus greening disease is a major cause of crop and tree loss in many parts of Asia and Africa. Before it was identified as one disease, it became known by various names: yellow shoot (huanglungbin) in China; likubin (decline) in Taiwan; dieback in India; leaf mottle in the Philippines; vein phloem degeneration in Indonesia; and yellow branch, blotchy-mottle, or greening in South Africa. As it became clear that all these were similar diseases the name "greening" was widely adopted. In India, this is one of the most important and widely prevalent mycoplasmal diseases of Citrus. Indian dieback was first accurately described in 1929 and attributed to poor drainage.

Causal organism : Mycoplasma Like Organism (MLO)

Vector

Graft Transmissibie and psyllid vectos, *Trioza erytreae*, *Diaphorina citri*.

Disease cycle

Greening has been experimentally transmitted by dodder to periwinkle (*Catharanthus roseus*) in which it induced marked yellowing. As in its citrus hosts, African greening in periwinkle required temperatures below 27°C for symptom development, whereas the Asian form was more heat-tolerant. The dodder itself is a host for the greening pathogen with titers higher than has been observed in citrus. Research on Asian greening followed close behind. In 1966 Salibe and Cortez demonstrated graft transmission in the Philippines, and reported on an insect vector, soon identified there and in India as another psylla, *Diaphorina citri*. The number of adult psylla of either species in a population that carry the disease is relatively small, but under experimental conditions, a single adult of either species can transmit greening. Other psylla species have been noted on citrus, namely *T. eastopi* (*T. litseae*) and Mesohomatoma lutheri in Réunion, *D. communis* in India. Organisms were observed in the hemolymph and salivary glands of *T. erytreae*, the phloem of yellow shoot and likubin-infected citrus and in infective *D. citri*.

Symptoms

Symptoms can occur throughout the tree, especially if the infection occurs at or soon after propagation. If infection occurs later the symptoms and the causal organism are often partially confined.

1. Infected trees or branches suffer heavy leaf drop followed by out-of-season flushing and blossoming, with dieback occurring in severe cases.
2. In general leaf symptoms are of two types: Primary symptoms are characterized by yellowing of normal-sized leaves along the veins and sometimes by the development of a blotchy-mottle. With secondary symptoms the leaves are small, upright, and show a variety of chlorotic patterns resembling those induced by zinc and iron deficiencies. Analysis of symptomatic leaves shows higher potassium content and lower calcium, magnesium, and zinc concentrations.
3. Infected fruit are small, lopsided and have a bitter taste, probably because of higher acidity and lower sugars. Many fall prematurely, while those that remain on the tree do not color properly, remaining green on the shaded side, hence the name of the disease. Any seeds in severely affected fruit are often abortive.
4. Root systems are poorly developed with relatively few fibrous roots, possibly because of root starvation. New root growth is suppressed and the roots often start decaying from the rootlets.
5. Amino acid concentrations are generally lower except for L-proline, which is higher in the leaves and flower bud. There

are also increases in amylase activity and glucose and fructose production, increased respiration early in infection and the appearance of gentisoyl-β-D-glucose, which seems to inhibit peroxidase activity. These changes do not, however, appear to be involved in symptom production.

6. Cytopathic studies have revealed pockets of necrotic phloem, excessive phloem formation, abnormal cambial activity and accumulation of starch in plasmids. Cytoplasmic membranes formed from in vaginations of the plasmalemma, aberration of the chloroplast thylakoids and mitochondrial collapse have also been observed. These changes indicate significant metabolic disturbances, possibly via toxins and/or hormones.

Control measures

The following control measures can be used for controlling the disease.

Thermotherapy

In 1964 Lin reported eliminating yellow shoot disease by water-saturated hot air treatment of graftwood at 48-58°C with no loss of tissue viability. Other attempts to eliminate Asian greening with heat have had varying degrees of success. In India treatment of budwood at 47°C for 2 h reduced disease incidence and longer treatments eliminated the pathogen. Treatment of infected young plants or seedlings budded with infected tissue at 38-40°C for three to four weeks also killed the pathogen. In South Africa budwood from infected trees heated over a hot water bath at 51°C for 1 h, 49°C for 2 h and 47°C for 4 h eliminated the disease although some tissue viability was lost at the higher temperatures. Infected trees covered for 2-5 months with polyethylene-covered fiberglass sheets showed a dramatic decrease in the number of diseased fruit. However this method is impractical for large-scale use. Heat treatments have some application in the elimination of greening from horticulturally desirable trees, possibly in conjunction with shoot tip grafting.

Chemotherapy

The association of prokaryotic organisms with greening prompted investigations into the use of antibiotics. Martinez et al. in the Philippines suppressed leaf symptoms with a foliar spray of tetracycline hydrochloride and apparently eliminated greening in budwood by immersion in the solution. In India immersion in a penicillin-carbendazin dip reportedly gave complete control. In South Africa Schwarz and van Vuuren injected various tetracyclines into the trunks of greening-infected sweet orange trees and found that tetracycline hydrochloride gave the best results, reducing fruit symptoms in the next crop from 60 to 20%. The incidence of infection remained unchanged for three years when a severe psylla outbreak and a slight increase in greening occurred. The trees received a second treatment that reduced fruit symptom incidence to below 10%. Better results were achieved when two injections were given one month apart, while a 97% reduction of fruit symptoms was achieved when trees were injected continuously under pressure for 7 days. The best time for injection is spring and distribution in the tree can be enhanced by the use of hyaluronidase. Trunk injections of tetracycline hydrochloride have also been successful in Taiwan, China, Réunion and the Philippines. Other antibiotics have been tested. Penicillin had a less suppressive effect than tetracycline hydrochloride in Asia and Réunion, but was totally ineffective in South Africa. In India partial to total remission occurred following injection with a compound called BP-101 (Hindustan Antibiotics Ltd.), ledermycin (demeclocycline hydrochloride) and a streptomycin-chlortetracycline mixture, while complete remission using foliar sprays of agrimycin and carbendazin has also been reported.

However, tetracycline hydrochloride is phytotoxic. Buds immersed in concentrations of over 250 ppm did not survive after grafting. Van Vuuren observed leaf narrowing, lamina yellowing, veinal browning, and occasional defoliation of individual twigs. A brown discoloration was found in the trunk in a zone extending some 300 mm from the hole, decreasing in intensity with distance. Leaf narrowing has also been recorded in Taiwan with oxytetracycline, but not with tetracycline hydrochloride or chlortetracycline.

A derivative of tetracycline hydrochloride, N-pyrrolidinomethyl tetracycline (PMT) is more soluble in water, giving slightly better control of greening and causing no foliar phytotoxicity and minimal wood discoloration. Midsummer treatments gave the best results. The solubility of PMT in water is 1250 mg/ml, compared to 10.8 mg/ml for tetracycline hydrochloride, so that smaller volumes can be injected from syringes without employing high pressures.

Moll was unable to detect by fluorimetry any antibiotic residues in the juice of injected trees one month after treatment. Other studies using growth inhibition of *Bacillus megaterium*, *B. cereus* var. *mycoides* or *B. subtilis*, have confirmed that antibiotic residues drop rapidly.

Use of Insecticides

Because psylla are sap feeders, systemic insecticides are the most effective. Those most commonly used in southern Africa are endosulfan sprays; monocrotophos applied as sprays, aerial sprays, trunk applications and trunk injections; and dimethoate applied to the soil. Weekly sprays of monocrotophos did not prevent infection, while high volume and mist blower applications destroy natural enemies of other citrus pests. A single application of dimethoate by microject and drip irrigation gave control for over 7 weeks, and two applications gave season-long protection. Simultaneous injection of dimethoate and tetracycline hydrochloride controlled psylla without reducing antibiotic efficacy. In India, *D. çitri* has been controlled for up to 4 weeks with sprays of several insecticides, including dimethoate and monocrotophos and soil applications of dimethoate. In China field control with soil applications of O-methoate has been reported. However insecticides can interfere with biological control of other pests. While dimethoate can reduce the efficacy of the biological of citrus red scale, at 0.01% a.i. it does not seriously affect the activity of psylla parasites.

Biological control

The major parasite of T. erytreae in South Africa is a species of parasitoid wasp Tetrastichus that oviposits in the psylla nymphs and appears to limit psylla populations. A survey in Réunion in 1973 found no parasitized mummies of psylla and it was suggested that parasitic wasps should be introduced. Three species, Tetrastichus radiatus (now called *Tamarixia radiata*) from India and *Tetrastichus dryi* and *Psyllaephagus pulvinatus* from South Africa, were collected and mass reared. *T. dryi* was released in 1976, and had dramatically reduced the *T. erytreae* population by 1978, with complete eradication by 1982. *P. pulvinatus* failed to establish itself. *T. radiata*, which is specific for *D. citri*, was introduced in 1978 and by 1982 had virtually eliminated psylla from commercial orchards, surviving only on *M. paniculata* hedges. The incidence of greening has consequently been reduced. Both species of wasp were subsequently introduced into Mauritius where they reduced psylla populations, but more slowly than in Réunion, possibly because they were not mass reared before release. *T. radiata* has also been introduced into Taiwan, the Philippines, Nepal and Indonesia, but without much apparent success. Other parasitic wasps of *D. citri* are *Diaphorencyrtus aligarhensis*, *Psyllaephagus* sp., *Chartocerus walkeri*, and *Encarsia*, but none appears to be a potential biological control agent. The efficacy of parasitic wasps is limited by the activity of hyperparasitic wasps. In South Africa the effect of Aphidencyrtus cassatus varies according to its ability to maintain synchrony with its host. Another species is *Cheiloneurus cyanonotus*. In Taiwan *D. aligarhensis* has at least 10 hyperparasites, of which one, Pachyneuron, can also oviposit in *T. radiata*. In China a species of *Tetrastichus* attacks both *T. radiata* and *D. aligarhensis*. No hyperparasites occur in Réunion, which may in part explain the success of biological control there. Fungi provide another means of biological control. *T. erytreae* is attacked by *Cladosporium oxysporium* and *Capnodium citri*, but both are sensitive to desiccation and are density-dependent. *D. citri* is parasitized by *Beauveria* and *Cephalosporiurn lacanii*, the latter somewhat effective at high density.

Other control measures

The production of greening-free trees is vital. The use of shoot-tip grafted material can ensure this goal. Using sticky yellow traps to identify nursery and orchard sites that are less susceptible to psylla attack can facilitate production. Such sites should also be distant from indigenous psylla hosts. Although psylla were considered not to have strong dispersal powers, recent investigation established that adult *T. erytreae* can be wind-dispersed in the absence of host plants 1.5 krn or more, and can survive approximately 85 hr without food. Strong winds such as those associated with typhoons have transported *D. citri* over medium to long distances. The removal of infected branches or trees and neglected trees may also reduce inoculum sources although mean population densities in the latter are not always significantly higher. If winter irrigation is withheld from lemon trees to suppress new flush, such infection reservoirs can be reduced. In China hedges of *M. paniculata*, which serve as hosts for *D. citri*, are being removed.

Sesamum Phyllody Disease

This disease is quite common in sesamum growing areas. Though the disease has been known since a long time, Pal and

Pushkarnath (1935) first reported it as a graft transmitted virus disease.

Causal organism : Mycoplasma Like Organism (MLO)

Vector : Graft Transmissible and *Orosius albicinctus*.

Disease cycle

This disease is transmitted by Orosius albicinctus. The incubation period of the pathogen in leafhoppers was found to be 15-63 days and 13-61 days in the host (*Sesamum*) depending upon the atmospheric conditions, especially the temperature. Both male and female leafhoppers were found to be equally infective. The disease could be transmitted through vectors or grafting to a number of plants of field crops, ornamentals and common weeds. Studies on the biology of the vectors showed that they can over winter at all stages and are present in nature throughout the year. The pathogen has a very wide host range. The important hosts are *Brassica Campestris* L. vars., toria, sarson, *Brassica rapa* and gram.

Symptoms

1. Vein clearing of the leaves, phylloid flowers, stimulation of axillary buds, which results in profuse branching, reduction in leaf size.
2. In affected plants, floral parts are transformed into green, leaflike structures resulting in complete sterility.
3. Shortening of internodes are some of the characteristics symptoms of the disease.

Control measures

The diseases are not important enough to merit any particular control measures. Late sowing of the crop may reduce in the incidence of the disease.

Rice Yellow Dwarf Disease

This disease was first observed in Japan in 1910. Now it is widespread in many South-east Asian countries. In India this disease has been reported from many parts of Orissa, Bihar, West Bengal, Andhra Pradesh and Delhi.

Causal organism : Mycoplasma Like Organism (MLO)

Vector : Green hoppers of rice *Nephotettix* spp.

Disease cycle

The disease is known to be transmitted by green hoppers of rice *Nephotettix* spp. The causal organism of this disease has been found to be associated in the phloem cells of the infected leaf tissues.

Symptoms

1. The disease is characterized by pronounced stunting and tillering with uniformly pale green to pale yellow chlorotic leaves.
2. Early infected plants die prematurely, while the late infected plants may survive but remain sterile. Sometimes they may bear a few poorly developed flowers and grains.

Control measures

Attempts have been made in japan to control the vectors of this disease by aerial application of insecticides over large areas. Seed dressing with systemic insecticides, Carbofuran and Propoxur can protect rice seedlings from the rice green leaf hoppers, the common vectors for both virus and rice yellow dwarf (Mitra et al., 1970).

Elm Yellows (Phloem Necrosis)

Causal organism : Phytoplasma

Vector : Phloem-feeding insects as Leafhoppers

Disease cycle

Elm yellows is caused by a phytoplasma (formerly called a mycoplasma-like organism). The disease is believed to be spread by such phloem-feeding insects as leafhoppers. The phytoplasma overwinters in infected tree roots and witches' brooms.

Symptoms

Symptoms of elm yellows may appear any time during the summer but are most common in mid- to late summer.

1. Symptoms of elm yellows include yellowing and drooping of foliage followed by leaf drop and death of branches.
2. This pattern may occur on one or a few branches or may quickly involve the entire tree.
3. Susceptible trees may show symptoms over the entire tree in a matter of a few weeks. Tolerant trees become stunted and may develop bunchy, prolific growth at the tips of branches (called witches' brooms) or on the trunk.
4. The inner bark tissues of infected trees often exhibit a butterscotch or light brown discoloration in small streaks or flecks.

Control measures

The diseases are not important enough to merit any particular control measures.

Witches'-broom Disease of Bean

Witches'-broom is caused by a species of mycoplasma but is of little economic importance in beans. However, it is worth mentioning because it represents a totally different category of disease.

Causal organism : Mycoplasmas

Vector : Leafhoppers and dodder plants.

Disease cycle

Mycoplasmas are harbored in perennial and biennial weeds and in eggs of infected leafhoppers in which they can propagate, passing through the eggs. Leafhoppers are the most important vectors that transmit the disease, but there are other means of disease transmission, namely by grafting scions and by parasitic dodder plants.

Symptoms

1. On beans, the infected plant is dwarfed and the internode shortened.
2. The infected plants produce extra shoots, and thus the plants are shooty and leafy.
3. Petioles are borne at wide angles, and flower peduncles are often elongated and wavy.
4. The leaves may curl slightly and later become somewhat yellow.
5. Pod sets are largely reduced. However, death seldom occurs in the beans.

Control measures

Control of leafhoppers by insecticides and removal of perennial and biennial weeds is the key to avoiding this disease.

Witches' Broom Diseases of Red Clover

Causal organism : Mycoplasma-like organism (MLO)

Vector : Leafhoppers

Disease cycle

The causal organism is transmitted by some kinds of leafhoppers.

Fig. 6.13. Witches' broom diseases of Red clover.

Symptoms

1. Mycoplasm-like organism disease causing shrinkage of the plant.
2. The disease mainly occurs in red clover. A lot of small leaves at first appear from the center part of the plant.
3. These leaves turn to reddish brown or fade.
4. The plant gradually begins to rosette and wither.

Control measures

Control of leafhoppers by insecticides and removal of perennial and biennial weeds is the key to avoiding this disease.

Stunt Disease of Bushes

Causal organism : Mycoplasma-like organism (MLO)

Vector : Leafhopper *Scaphytopius magdalensis*

Disease cycle

Stunt is a serious disease caused by a mycoplasma-like organism (MLO). Almost all cultivars are susceptible to stunt. Only Rancocas cultivar has resistance. The sharpnosed leafhopper (*Scaphytopius magdalensis*) spreads this disease.

Symptoms

1. The disease causes a stunting of bushes.
2. Leaf symptoms are the best evidence for disease diagnosis. Leaves show downward cupping and yellowish margins.
3. Normal green coloration of the leaf is usually retained along the midvein and lateral veins. Leaf size is usually reduced.
4. Berries are smaller than normal on infected bushes.

Control measures

Remove and burn infected bushes. The normal full-season insecticidal control program for blueberries usually keeps the leafhopper population in check. Replant with disease-free stock.

Prunus X Disease

This is also known as Eastern peach X disease, Western peach X disease.

Causal organism : X Disease Phytoplasma

Vector : Leafhoppers

Symptoms

1. Symptoms appear six weeks after foliation. Blotches develop across the leaf blades; first appearing water-soaked then turning red/yellow.
2. The rest of the leaf becomes chlorotic with margins rolled upward. Leaves become tattered as the blotches drop out.
3. Affected leaves fall early, leaving only small rosetted leaf tufts at the shoot tips. In early stages, symptoms are limited to few branches; the disease then progresses randomly.
4. Most fruits drop soon after the first symptoms develop. Those remaining are angular and color early. They produce shriveled, nonviable seeds, and discolored flesh around the pits.

Control measures

The diseases are not important enough to merit any particular control measures.

Peach Yellows Disease

This is also known as Peach virus 1, Prunus virus 1, Chlorogenus persicae caused disease.

Causal organism : X Disease Phytoplasma

Vector : Leafhoppers

Symptoms

1. Leaves are chlorotic, yellow, and often roll upward; red spotting develops as leaves begin to droop.
2. Severe infections will develop slender shoots with small, narrow, yellow leaves.
3. Buds that are normally latent on younger trees produce dwarfed leaves.
4. Terminal dieback is common in one-year-old shoots. Trees usually die 2 to 3 years following the appearance of symptoms.
5. Fruits are of poor quality and tend to reach maturity three weeks earlier than normal.

6. Red-colored cultivars have saturated coloration, which may be spotted. The flesh exhibits red streaks or marble patterns, with redness around the pit.

Control measures

The diseases are not important enough to merit any particular control measures.

Peach Vein Clearing Disease

A study was conducted from 1982 to 1990 in the Pyrenees-Orientales to investigate susceptibility of the Rouge du Roussillon apricot cultivar to apricot chlorotic leaf roll phytoplasma diseases (ACLR). The trees in the study were found to have the following atypical symptoms, and have been identified as peach vein clearing.

Causal organism : Unknown phytoplasma.

Symptoms

1. In summer, mild chlorosis and partial rolling of the leaves were apparent while the veins seemed to be juttying.
2. Vegetation is prolonged in autumn.
3. The number of superior sized fruit is reduced. Trees do not decline.
4. In winter, bark became swollen and bud burst was delayed from 6 to 8 days.

Control measures

The diseases are not important enough to merit any particular control measures.

Peach Rosette Disease

This is also known as Peach rosette virus, Peach virus 2, Prunus virus 2, Carpophthora rosettae, Nanus rosettae, Persicavirus rosettans caused diseases.

Causal organism : X disease strain phytoplasma.

Symptoms

1. Initial symptoms develop quickly. The first leaves reach normal size but become yellow and roll inward or arch backward, their veins sometimes appearing red. Defoliation of these leaves follows.
2. Internodes of new shoots are extremely short with leaves pressed into dense rosettes.

3. As leaves on older parts of the trees have already fallen, only tufts of young leaves on the tips of bare shoots remain.
4. Blossoms rarely develop and/or the flowers do not set fruit on affected branches.
5. Most diseased trees die within the first year of infection; those that leaf out the following spring die shortly thereafter.

Control measures

The diseases are not important enough to merit any particular control measures.

Peach Red Suture Disease

This is also known as Peach virus 5, Prunus virus 4, Persicavirus rubescens caused disease.

Causal organism : X disease phytoplasma.

Symptoms

Symptoms caused by the disease are generally attributed to a strain of peach yellows phytoplasma.

1. Fruits of affected trees mature and soften several days earlier than normal on their suture sides, while their converse sides remains green and hard.
2. The suture side often becomes swollen; its flesh, coarse and watery.
3. Fruits may develop irregular contours, and in red-fruited cultivars, the suture side remains a deeper red with purple areas at the apexes.
4. Leaves exhibit a yellowish-green or bronze color, noticeable after petal fall and before harvest.
5. Foliage also develops autumnal coloring earlier than normal.
6. On two-year-old branches, buds may grow into spur-like outgrowths.

Control measures

The diseases are not important enough to merit any particular control measures.

Peach Little Peach Disease

This is also known as Peach virus 3, Prunus virus 1A, Chlorogenus persicae var. micropersica caused disease.

Causal organism : X disease strain phytoplasma.

Symptoms

1. Leaves of infected trees are distinctly greener than normal during the early part of the vegetation period.
2. Proliferation on side branches and spurs of 2 to 3-year-old branches gives the tree a bushy stature.
3. Shoots are long and upright.
4. Leaves are leathery and droop downwards, later becoming chloroitic.
5. Symptoms first appear on one part of the tree, and then spread through its entirety by the third year.
6. Maturation of the small and malformed fruits is delayed, sometimes as much as by 3 weeks.
7. Their pits are small, their kernels underdeveloped.
8. Flavor of affected fruits is insipid.

Control measures

The diseases are not important enough to merit any particular control measures.

Molieres Disease

A rare disease affecting sweet cherry and European plum, Molieres disease appears to be restricted to southwestern France.

Causal organism : European Stone Fruit Yellows Phytoplasma.

Symptoms

1. Initial symptoms appear as slightly chlorotic leaves in summer.
2. The following year the trees bloom abundantly, though flowers are malformed and fruit set is poor.
3. The few fruit that do develop remain small and drop prematurely.
4. Small deformed leaves, rosetting and poor lignification of young shoots as well as phloem and bark necrosis also are associated with the disease.
5. Young shoots die back and as the disease progresses the entire scaffolds die.
6. In new orchards, diseased trees are usually not observed until 3 to 7 years after planting.
7. All cherry cultivars and rootstocks, especially *P. mahaleb*, appear to be susceptible, however, plum cultivars differ in susceptibility.

Control measures

The diseases are not important enough to merit any particular control measures.

Cherry Decline Disease

Causal organism : Unknown phytoplasma.

Symptoms

Symptoms are attributed to a complex of the Prune dwarf ilavirus and Strawberry latent ringspot nepovirus, which induce Cherry Rasp Leaf (European).

1. Enations, in the form of leafy outgrowths or small protuberances along the midrib, appear on the lower surfaces of leaves.
2. In other cases, only chlorotic areas ("oil flecks") are visible on the leaves. Additional symptoms may include narrow, folded leaves; bud abortion; twig dieback; or terminal tufts of leaves.
3. Fruits of black cherry develop red spots as the yields decline.
4. Necrotic symptoms appear on secondary roots, and cankerous wounds and gummosis develops on branches.

Control measures

The diseases are not important enough to merit any particular control measures.

Cherry Blossom Anomaly Diseases

Causal organism : X Disease Phytoplasma.

Symptoms

1. Leaf symptoms are similar to those of boron deficiency as the boron content of diseased tissues is low. Leaves are elongated and narrow and curve inward. This may be accompanied by interveinal chlorosis. In older infections, leaves originate from terminal buds and form rosettes at the tips of bare branches.
2. Normally white petals may exhibit varied amounts of green coloration starting at the midrib. This usually shows as stripes, but may encompass the entire petal. Some petals appear leaf like and some flower buds look like leaf buds, except for their yellow anthers.
3. Fruits developing from affected blossoms tend to be pointed, have roughened skin, and usually have a persistent style.

Control measures

The diseases are not important enough to merit any particular control measures.

Cherry Albino Disease

Causal organism : X Disease Phytoplasma.

Symptoms

Nearly the same symptoms appear in all sweet cherries, differences occur only in the sensitivities of each cultivar.

1. Leaves roll inward and turn a golden bronze or olive-brown in color as early as June.
2. Coloring appears mostly along the main or secondary veins to form feathery patterns.
3. A general symptom of the disease has the leaves produced in late summer from terminal buds forming rosettes.
4. Fruits remain small and fail to mature, assume a light color, and drop prematurely. In some cases, small white fruits remain on the tree longer.

Control measures

The diseases are not important enough to merit any particular control measures.

Apricot Chlorotic Leaf Roll Disease

Causal organism : European stone fruit yellows phytoplasma.

Fig. 6.14. Apricot chlorotic leaf roll disease.

Symptoms

Symptoms of this disease vary; young trees may become systematically infected, whereas in trees older than 5 years, initial symptoms are localized to lower branches and spread to the crown. In either case, infected trees do not live long.

1. Early bud opening and foliation are reliable symptoms.
2. Characteristic symptoms first appear in May as the blades roll inward along convergent lines and turn pale and chlorotic.
3. Leaf fall is delayed, and abnormal bud opening continues in freezing temperatures.
4. Fruits are undersized, sometimes bumpy, and drop prematurely. The flesh is brown and spongy near the pit.
5. Necrosis may develop in the bark, and is visible in transverse sections as either thin orange lines or thicker brown bands.

Control measures

The diseases are not important enough to merit any particular control measures.

Another Mycoplasmal Diseases

There are following some another mycoplasmal diseases are also found in different plants.

I. ***Witches' broom of Chinese jujube*** (*Ziziphus jujuba* Mill.) is found in China and Korea. The causal agent has been confirmed to be a mycoplasma-like organism (Zhang et al. 1991, Zhu et al. 1988). The MLO is transmitted by a leaf hopper, *Hishimonoides chinensis* Anufrive.

II. ***Witches' broom of longan*** (*Euphoria longan* Lam.) is found in China (Chen et al. 1992) and neighboring countries. A flexible virus has been isolated from infected longan trees. The virus is transmitted by a stinkbug, *Tessaratoma papillosa* Drury, and a psylliid, Cornegenapsylla sinica Yang et Li. It is transmitted from one longan tree to another, and also from longan to litchi (*Litchi chinensis Sonnerat*). The symptoms in longan are very similar to those produced by litchi witches' broom disease.

III. ***Marginal flavescence of potato disease*** is prevalent in hills and the plateau area where cool weather prevails for a longer time during the crop season. The loss due to this disease may vary from 40 to 75 per cent (Nagaich, et al., 1974). The symptoms of Marginal flavescence of potato disease are chlorosis on the margins of upper leaves. In rare cases the severely affected plants, after exhibiting flavescence of leaf margins for quite some time develop pinkish colouration of leaflet bases and start rolling. Hairy root symptoms on tubers of 60 to 65 percent plants affected with MF have been observed. The pathogen is leafhopper and graft transmitted. Five to 35 per cent tubers carry the pathogen and serve as a source of primary inoculum. Hot water treatment of tubers at 500^{0}C for 10 to 15 minutes eliminates the pathogen without adversely affecting the tubers (Sangar and Singh, 1985)

IV. ***Withes broom of potato disease*** results in extreme stunting of the plant and numerous filamentous stems with simple leaves. The affected plants resulting in total loss of the plant produce very small tubers. The pathogen is leafhopper transmitted and perennates through tubers.

REFERENCES

Abou-Jawdah Y, Karakashian A, Sobh H, Martini M, Lee IM. 2002. An epidemic of almond witches'-broom in Lebanon: classification and phylogenetic relationships of the associated phytoplasma. Plant Disease 86: 477-84

Adams AN, Davies DL, Kirby MJ. 2001. Virus and phytoplasma detection in fruit trees. Outlook on Agriculture 30: 45-54.

Agrios, G.N. (1997). Plant diseases caused by Mollicutes: phytoplasmas and spiroplasmas. In Plant Pathology, 4th, pp. 457-470. Edited by G. N. Agrios. New York: Academic Press.

Agrios, G.N. (1997). Plant Pathology, 4th ed. Academic Press, San Diego.

Ahmad I, Soomro MH, Khalid S, Iftikhar S, Munir A, Burney K. 1995. Recent distributional trends of potato diseases in Pakistan. Research and development of potato production in Pakistan. Proceedings of the National Seminar held at NARC, Islamabad, Pakistan

Ahmed W, Farooq A. 1994. Witches' broom; a new disease on potato in Punjab. Pakistan Journal of Phytopathology 6: 67-8.

Akerlund, T., K. Nordstrom, and R. Bernander. 1993. Branched Escherichia coli cells. Mol. Microbiol. 10:849-858.

Al-Ani RA, Al-Fadhil KF, Al-Rawi FA, Al-Shamri AN. 2001. Serological relationship between suaeda Witches' broom phytoplasma and those infecting some crops and weed plants in Iraq. Arab Journal of Plant Protection 19: 59-64.

Alma, A., Bosco, D., Danielli, A., Bertaccini, A., Vibrio, M., and Arzone, A. (1997). Identification of phytoplasmas in eggs, nymphs, and adults of Scaphoideus titanus Ball reared on healthy plants. Insect Mol. Biol. 6: 115-121.

Alma, A., R.E. Davis, M. Vibio, A. Danielli, D. Bosco, A. Arzone, and A. Bertaccini. 1996. Mixed infection of grapevines in northern Italy by phytoplasmas including 16S rRNA RFLP subgroup 16SrI-B strains previously unreported in this host. Plant Dis. 80: 418-421.

Bachand GD, Castello JD. 1998. Abstracts of the conference 'Phytoplasma and virus diseases of forest trees' (IUFRO-WP 07.04.02), College of Environmental Science and Forestry, State University of New York, Syracuse, NY, USA, June 11-13, 1997. European Journal of Forest Pathology 28: 75-83.

Bairoch, A. and Apweiler, R. (1998). The SWISS-PROT protein sequence data bank and its supplement TrEMBL in 1998. Nucl. Acids Res., 26, 38-42.

Barbara D.J., D.L. Davies and M.F. Clark (1998) - Cloning and sequencing of a major membrane protein from chlorante (aster yellows) phytoplasma. Abstr. G.04. 12th Intern. Organization Mycoplasmology, 22-28 July, 1998, Sydney.

Barloy, J. and J. Pelhate 1962. PremiËres observations phytopathologiques relatives aux cultures de chanvre en Anjou. Ann. Epiphyties 13: 117-149.

Baseman JB, Dallo SF, Tully JG, Rose DL. Isolation and characterization of Mycoplasma genitalium strains from the human respiratory tract. J Clin Microbiol 1988;26:2266-9.

Baseman JB, Lange M, Criscimagna NL, Girón JA, Thomas CA. Interplay between mycoplasmas and host target cells. Microb Pathog 1995;19:105-16.

Baseman JB, Reddy SP, Dallo SF. Interplay between mycoplasma surface proteins, airway cells, and the protean manifestations of mycoplasma-mediated human infections. Am J Respir Crit Care Med 1996;154:S137-44.

Baseman JB. The cytadhesins of Mycoplasma pneumoniae and M. genitalium. In: Rottem S, Kahane I, editors. Subcellular biochemistry. New York: Plenum Press, 1993;243-59.

Beardsley, J. W. 1970. Notes and Exhibitions. Proc. Hawaii Entomol. Soc. 20(3): 484-487.

Berg, M., D.L. Davies, M.F. Clark, J. Vetten, G. Maier and E. Seemuller. (1999) Isolation of a gene encoding an immunodominant membrane protein gene of the apple proliferation phytoplasma and expression and characterization of the gene product. Microbiology (in press).

Bernstein, F. C., Koetzle, T. F., Williams, G. J. B., Meyer, E. F., Brice, M. D. et al. (1977). The Protein Data Bank: a computer based archival file for macromolecular structures. J. Mol. Biol., 112, 535-542.

Bhagabati, K. N., Nariani, T. K. 1980. Interaction of greening and tristeza pathogens in Kagzi lime (Citrus anrantifolia (Christm.) Swing.) and their effect on growth and development of disease symptoms. Indian Phytopathol. 33:292-95.

Bhagabati, K. N., Nariani, T. K. 1983. Chemical control of citrus psylla, a vector of citrus greening disease. J. Res. Assam. Agric. Univ. 4:86-88 (Rev. Appl. Entomol. Ser. A 75:28).

Bhavakul, K., Intavimolsri, S., Vichitrananda, S., Kratureuk, C., Prommintara, M. 1981. The current citrus disease situation in Thailand with emphasis on citrus greening. Proc. Int. Soc. Citricult. 1:464-66.

Bianco, P.A., R.E. Davis, J.P. Prince, I.-M. Lee, D.E. Gunderson, A. Fortusini, and G. Belli. 1993. Double and single infections by aster yellows and elm yellows MLOs in grapevines with symptoms characteristic of flavescence doree. Riv. Pat. Veg., S.V. 3: 69-82.

Biberfeld G. Infection sequelae and autoimmune reactions in Mycoplasma pneumoniae infection. In: Razin S, Barile MF, editors. The Mycoplasmas Vol. IV. New York: Academic Press, 1985:293-311.

Biberfeld, G., and P. Biberfeld. 1970. Ultrastructural features of Mycoplasma pneumoniae. J. Bacteriol. 102:855-861.

Bilgrami, K., S. Jamaluddin and M. A. Rizwi 1981. Fungi of India: Part II, Host index and addenda. Today & Tomarrow's Printers and Publishers, New Delhi. 128 pp.

Bird, J., and Adsuar, J. 1952. Viral nature of papaya bunchy top. J. Agr. Univ. Puerto Rico 36:5-11.

Bisset LR. The Mycoplasma genitalium adhesin protein and several human class II MHC proteins exhibit sequence homology: possible ramifications for the development of autoimmunity. Autoimmunity 1992;14:167-8.

Blanchard A, Montagnier L. AIDS-associated mycoplasmas. Annu Rev Microbiol 1994;48:687-712.

Blattner, F. R., Plunkett, G., Bloch, C. A., Perna, N. T., Burland, V. et al. (1997). The complete genome sequence of Escherichia coli K-12. Science, 277, 1453-1474.

Boatman, E. S. 1979. Morphology and ultrastructure of the mycoplasmatales, p. 63-102. In M. F. Barile, and S. Razin (ed.), The mycoplasmas, vol. I. Academic Press, New York, N.Y.

Bork, P., C. Ouzounis, G. Casari, R. Schneider, C. Sander, M. Dolan, W. Gilbert, and P. M. Gillevet. 1995. Exploring the Mycoplasma capricolum genome: a minimal cell reveals its physiology. Mol. Microbiol. 16:955-967.

Borodina E. I. and N. D. Migal 1987. Flower teratology in intersexual hemp plants. Soviet Journal of Developmental Biology 17(4): 262-269.

Bosco, D., C. Minucci, G. Boccardo, M. Conti. 1997. Differential acquisition of chrysanthemum yellows phytoplasma by three leafhopper species. Entomologia Experimentalis et Applicata. 83:219-224.

Bové J.M., P. Carle, M. Garnier, F. Laigret, J. Renaudin and C. Saillard 1989. Molecular and cellular biology of spiroplasmas. p. 243-364. in R.F. Whitcomb & J.G. Tully (eds). The Mycoplasmas, vol. V, Academic Press, New York.

Bowyer, J.W., J.G. Atherton, D.S. Teakle, and G.A. Ahern. 1969. Mycoplasma-like bodies in plants affected by legume little leaf, tomato big bud, and lucerne witches' broom diseases. Aust. J. Biol. Sci. 22:271-274.

Brand, R.J. and M.B. von Wechmar. 1993. Characterization of two viruses implicated in the woodiness disease complex of South African passionfruit: Cucumber mosaic virus and a new potyvirus. Journal of the Southern African Society for Horticultural Science 3: 28-33.

Brannigan, J. A., Dodson, G., Duggleby, H. J., Moody, P. C. E., Smith, J. L. et al. 1995. A protein catalytic framework with an N-terminal nucleophile is capable of self-activation. Nature, 378, 416-419.

Brown, J.K., J. Bird and D.C. Fletcher. 1993. First report of passiflora leaf mottle disease caused by a whitefly-transmitted geminivirus in Puerto Rico. Plant Disease 77: 1264.

Buitendag, C. H., Bronkhorst, G. J. 1986. Further aspects of trunk treatment of citrus with insecticides: Phytotoxicity, side effects on incidental pests and development of application apparatus. Citrus Subtrop. Fruit J. 623:7-10.

Bult, C. J., White, O. W., Olsen, G. J., Zhou, L. Z., Fleischmann, R. D. et al. 1996. Complete genome sequence of the Methanogenic Archaeon, Methanococcus jannaschii. Science, 273, 1058-1073.

Burdge DR, Reid GD, Reeve CE, Robertson JA, Stemke GW, Bowie WR.1988. Septic arthritis due to dual infection with Mycoplasma hominis and Ureaplasma urealyticum. J Rheumatol ;15:366-8.

Burr, T. J., J. L. Norelli, B. Katz, W. F. Wilcox, and Hoying, S. A. 1988. Streptomycin resistance of Pseudomonas syringae pv. papulans in apple orchards and its association with a conjugative plasmid. Phytopathology 78:410-413.

Buxton, A., and G. Fraser. 1977. Mycoplasma, Acholeplasma and L-form of bacteria, p. 267-283. Animal microbiology, vol. 1. Blackwell Scientific Publications Ltd., Oxford, United Kingdom.

Capoor, S. P., Rao, D. G., Viswanath, S. M. 1967. Diaphorina citri Kuway., a vector of the greening disease of citrus in India. Indian J. Agric. Sci. 37:572-76.

Chagas, C.M., P.P. Joazeiro, M. Kudamatsu, and J. Vega. 1984. Mosaic of purple granadilla, a new virus in Brazil. Fitopatologia Brasileira 9: 241-247.

Chakraborty, N. K., Pandey, P. K., Chatterjee, S. N., Singh, A. B. 1976. Host preference in Diaphorina citri Kuwayama, vector of greening disease in India. Indian J. Entomol. 38:196-97 (Rev. Appl. Entomol. Ser. A 67:232).

Chang, C.A. 1992. Characterization and comparison of passionfruit mottle virus, a newly recognized potyvirus, with passionfruit woodiness virus. Phytopathology 82: 1358-1363.

Chang, C.A. and Y.D. Lin. 1992. Ecology of passionfruit virus transmission and the evaluation of the effectiveness of using virus-free seedlings to control passionfruit virus diseases in Taiwan. Plant Pathology Bulletin, Taiwan Agricultural Research Institute 1: 140-146.

Chang, C.J. 1998. Pathogenicity of aster yellows phytoplasma and Spiroplasma citri on periwinkle. Phytopathology 88:1347-1350.

Cheema, S. S., Kapur, S. P., Bansal, R. D. 1985. Efficacy of various therapeutic agents against greening disease of citrus. J. Res. Punjab Agric. Univ. 22:479-82 (Rev. Plant Pathol. 65:551).

Chen, J.Y., C.F. Xu, K.B. Li, and Y.H. Xia. 1992. On transmission of longan witches' broom disease by insect vectors. Acta Phytopathologica Sinica 22: 245-249.

Chen, K.H., J.R. Guo, X.Y. Wu, N. Loi, L. Carraro, Y.H. Guo, Y.D. Chen, R. Osler, R. Pearson, and T.A. Chen. 1993. Comparison of monoclonal antibodies, DNA probes, and PCR for detection of the grapevine yellow disease agent. Phytopathology. 83:915-922.

Chen, T.A., and Y.P. Jiang 1990. Progress in the detection of plant MLO by using Monoclonal and polyclonal antibodies, pp. 270-275. In G. Stanek, G.H. Cassell, J.G. Tully, and R.F. Whitcomb (eds.), Recent advances in mycoplasmology. Proceedings of the 7th congress of the International Organization for Mycoplasmology, Baden near Vienna, 1988. Verlag: New York.

Chen, T.A., J. M. Wells, C. H. Liao 1982. Cultivation in vitro: spiroplasmas, plant mycoplasmas, and other fastidious, walled prokaryotes, pp. 417-446. In M.S. Mount, Lacy, G. H. (ed.), Phytopathogenic Prokaryotes, vol. 2, Associated Press.

Chen, T.A., J.D. Lei, and C.P. Lin. 1989. Detection and identification of plant and insect mollicutes. Pages 393-424 in The Mycoplasmas Volume V: Spiroplasmas, Acholeplasmas, and Mycoplasmas of

Plants and Arthropods, R.F. Whitcomb and J.G. Tully, eds. Academic Press Inc., San Diego, CA. 653 p.

Chen, Y.D. and T.A. Chen. (1998) Expression of engineered antibodies in plants: a possible tool for spiroplasma and phytoplasma disease control. Phytopathology. 88:1367-1371.

Chingbingyong MI, Hughes CV. 1996. Detection of Mycoplasma fermentans in human saliva with a polymerase chain reaction-based assay. Archs Oral Biol 41:311-4.

Chiou, C.-S., and A. L. Jones. 1993. Nucleotide sequence analysis of a transposon (Tn5393) carrying streptomycin resistance genes in Erwinia amylovora and other gram-negative bacteria. J. Bacteriol. 175:732-740.

Chiou, C.-S., and A. L. Jones. 1995a. Expression and identification of the strA-strB gene pair from streptomycin-resistant Erwinia amylovora. Gene 152:47-51.

Chiykowski, L. N., M.T. Cousin, J.L. Dale, G.T.N. de Leeuw, D.A. Golino, K.J. Hackett, B.C. Kirkpatrick, R. Marwitz, H. Petzold, R.C. Sinha, M. Sugiura, R.F. Whitcomb, I.L. Yang, B.M. Zhu, and E. Seemuller 1989. Plant Diseases associated with mycoplasma-like organisms., pp. 546-640, The Mycoplasmas, vol. V, Academic Press.

Chiykowski, L.N. 1988. Maintenance of yellows-type mycoplasmalike organisms. Pages 123-134 in Tree Mycoplasmas and Mycoplasma Diseases, C. Hiruki, ed., The University of Alberta Press, Edmonton, AB. 245 pp.

Chiykowski, L.N. 1991. Reaction of additional barley cultivars to two aster yellows strains. Can. Plant Dis. Surv. 71: 143-145.

Chothia, C. 1992. One thousand protein families for the molecular biologist. Nature, 357, 543-544.

Clark, M.F., D.L. Davies, N. Boonham, S.J. Crossley, G. Keane and D.J. Barbara. 1996. Cloning membrane-associated protein genes from an aster yellows phytoplasma and comparisons with the gene sequences from other phytoplasmas. IOM Lett. 4:267.

Colariccio, A., C.M. Chagas, M. Mizuki, J. Vega, and E. Cereda. 1987. Natural infection of golden passion fruit with cucumber mosaic virus in Brazil. Fitopatologia Brasileira 12: 254-257.

Cole BC. Mycoplasma interactions with the immune system: implications for disease pathology. ASM News 1996;62:471-5.

Cole, R. M. 1983. Transmission electron microscopy: basic techniques, p. 43-50. In S. Razin, and J. S. Tully (ed.), Methods in mycoplasmology. Academic Press, New York, N.Y.

Coyier, D. L., and R. P. Covey. 1975. Tolerance of Erwinia amylovora to streptomycin sulfate in Oregon and Washington. Plant Dis. Rep. 59:849-852.

Crestani, O.A., E.W. Kitajima, M.T. Lin, and V.L.A. Marinho. 1986. Passion fruit yellow mosaic virus, a new tymovirus found in Brazil. Phytopathology 76: 951-955.

Cunningham MW. Molecular mimicry: bacterial antigen mimicry. In: Bona CA, Siminovitch K, Theofilopoulos AN, Zanetti M, editors. 1993. The pathology of autoimmunity. New York: Harwood Academic Publishers, 245-56.

Dallo SF, Baseman JB. Adhesin gene of Mycoplasma genitalium exists as multiple copies. Microb Pathog 1991;10:475-80.

Dallo SF, Chavoya A, Baseman JB. Characterization of the gene for a 30-kilodalton adhesin-related protein of Mycoplasma pneumoniae. Infect Immun 1990;58:4163-5.

Dallo SF, Horton JR, Su CJ, Baseman JB. Restriction fragment length polymorphism in the cytadhesin P1 gene of human clinical isolates of Mycoplasma pneumoniae. Infect Immun 1990;58:2017-20.

Dallo SF, Su CJ, Horton JR, Baseman JB. Identification of P1 gene domain containing epitope(s) mediating Mycoplasma pneumoniae cytadherence. J Exp Med 1988;167:718-23.

Dassanayake, E.M. and R.G.T. Hicks. 1992. Sri Lanka passion fruit mottle virus, a potyvirus infecting golden passion fruit in Sri Lanka. Annals of applied Biology 120: 459-469.

Davis, M. J., J. B. Kramer, F. H., Ferwerda, and B. R. Brunner. 1996. Association of a bacterium and not a phytoplasma with papaya bunchy top disease. Phytopathology 86:102-109.

Davis, M.J., J.B. Kramer, F.H. Ferwerda, and B.R. Brunner. 1996. Association of a bacterium and not a phytoplasma with papaya bunchy top disease. Phytopathology. 86:102-109.

Davis, R. E. and I-M. Lee. 1993. Cluster-specific polymerase chain reaction amplification of 16S rDNA sequences for detection and identification of mycoplasmalike organisms. Phytopathology 83: 1008-1011.

Davis, R. E., Jomantiene, R. and E. L. Dally. 1998. Interoperon sequence heterogeneity and differential PCR-mediated amplification of sequences from the two rRNA operons in phytoplasmas. IOM Letters 5: 176-177.

Davis, R.E. 1981. The enigma of the flower phytoplasmas. Pages 259-279. In, Mycoplasma Diseases of Trees and Shrubs. Academic Press, New York.

Davis, R.E., and Sinclair, W.A. (1998) Phytoplasma identity and disease etiology. Phytopathology 88:1372-1376.

Davis, R.E., Dally, E.L., Gundersen, D.E., Lee, I.-M., and Habili, N. (1997) "Candidatus Phytoplasma australiense," a new phytoplasma taxon associated with Australian grapevine yellows. Int. J. Syst. Bacteriol. 47:262-269.

De Wijs, J.J. 1974. A virus causing ringspot of Passiflora edulis in the Ivory Coast. Annals of applied Biology 77: 33-40.

Deeley, J., W.A. Stevens, and R.T.V. Fox. 1979. Use of Dienes' Stain to detect plant diseases induced by MLOs. Phytopathology. 69:1169-1171.

Deguichi T, Komeda H, Yasuda M, Tada K, Iwata H, Asano M, et al. Mycoplasma genitalium in non-gonococcal urethritis. Int J STD & AIDS 1995;6:144-6.

Denes, A. S. and Sinha, R. C. 1991. Extrachromosomal DNA elements of plant pathogenic mycoplasmalike organisms. Can J Plant Pathol 13, 26-32.

Dinesh M R et al. 1991. Combining ability studies in Carica papaya L. with respect to yield and quality characters. GARTENBAU-WISSENSCHAFT (GERMANY). v. 56(2) p. 81-83. (02237)

Doi, Y., Teranaka, M., Yora, K. and H. Asuyama. 1967. Mycoplasma or PLT-group-like microorganisms found in the phloem elements of plants infected with mulberry dwarf, potato witches-broom, aster yellows, or paulownia witches-broom. Annals of the Phytopathological Society of Japan 33: 259-266.

Drake, D. C. and Chapman, R.K. 1965. Evidence for long distance migration of the six-spotted leafhopper into Wisconsin. In Migration of the six-spotted leafhopper Macrosteles fascifrons (Stal). University of Wisconsin Res. Bull. 261.

Drexler, Hans G. and Cord C. Uphoff. 2002. Mycoplasma contamination of cell cultures: Incidence, sources, effects, detection, elimination, prevention. Cytotechnology, 39: 75-90.

Dybvig, K., and Voelker, L.L.. 1996. Molecular biology of mycoplasmas. Annu. Rev. Microbiol. 50:25-57.

Ebesu, R. H. 1985. The Biology of the Leafhopper Empoasca stevensi Young (Homoptera: Cicadellidae) and its Toxicity to Papaya. MS. Thesis. University of Hawaii, Department of Entomology.

Eden-Green, S.J. 1982. Culture of other microorganisms from yellows-diseased plants, pp. 201-239. In M.J.D.a.P.G. Markham (ed.), Plant and Insect mycoplasma techniques. Croom and Helm, London.

El-Shafy, A., A. Fudl-Allah, E. C. Calavan, E. C. K. Igwegbe. (1972). Culture of a MLO assoc. with stubborn disease of citrus. Phytopathology. 62:729-731.

Errampalli, D., J. Fletcher, and P.L. Claypool. 1991. Incidence of yellows in carrot and lettuce and characterization of mycoplasmalike organism isolates in Oklahoma. Plant Dis. 75: 579-584.

Firrao G., C.D. Smart and B.C. Kirkpatrick 1996. Physical map of the western X-disease phytoplasma chromosome. J. Bacteriol., 178 (13), 3985-3988.

Fischer, D. and Eisenberg, D. 1997. Assigning folds to the proteins encoded by the genome of Mycoplasma genitalium. Proc. Natl. Acad. Sc. U.S.A., 94, 11929-11934.

Fletcher, J., A. Wayadande, U. Melcher, and F. Ye. 1998. The phytopathogenic mollicute-insect vector interface: a closer look. Phytopathology 88:1351-1358.

Foissac X., C. Saillard and J.M. Bové 1997. Random insertion of transposon Tn4001 in the genome of Spiroplasma citri strain GII3. Plasmid, 37, 80-86.

Foissac X., J.L. Danet, C. Saillard, P. Gaurivaud, F. Laigret, C. Paré and J.M. Bové1997. Mutagenesis by insertion of Tn4001 into the genome of Spiroplasma citri: characterization of mutants affected in plant pathogenicity and transmission to the plant by the leafhopper vector Circulifer haematoceps. Mol. Plant-Microbe Interact., 10 (4), 454-461.

Foissac, X., J.L. Danet, C. Saillard, P. Gaurivaud, F. Laigret, C. Pare, and J.M. Bove. 1997. Mutagenesis by insertion of Tn4001 into the genome of Spiroplasma citri: Characterization of mutants affected in plant pathogenicity and transmission to the plant by the leafhopper vector Circulifer haematoceps. MPMI 10:454-461.

Fokkema, N. J. and J .Van Den Heuvel 1986. Microbiology of the phyllosphere. Cambridge University Press, Cambridge. 392 pp.

Fox, G.E., Wisotzkey, J.D., and Jurtshuk, P., Jr. 1992. How close is close: 16S rRNA sequence identity may not be sufficient to guarantee species identity. Int. J. Syst. Bacteriol. 42:166-170.

Fraser, C. M., Casjens, S., Huang, W. M., Sutton, G. G., Clayton, R. et al. (1997). Genomic sequence of a Lyme disease spirochaete, Borrelia burgdorferi. Nature, 390, 580-586.

Fraser, C. M., J. D. Gocayne, O. White, M. D. Adams, R. A. Clayton, R. D. Fleischmann, C. J. Bult, A. R. Kerlavage, G. Sutton, J. M. Kelley, R. D. Fritchman, J. F. Weidman, K. V. Small, M. Sandusky, J. Fuhrmann, D. Nguyen, R. Utterback, D. M. Saudek, C. A. Phillips, J. M. Merrick, J.-F. Tomb, B. A. Dougherty, K. F. Bott, P.-C. Hu, T. S. Lucier, S. N. Peterson, H. O. Smith, C. A. Hutchison III, and J. C. Venter. 1995. The minimal gene complement of Mycoplasma genitalium. Science 270:397-403.

Furr PM, Taylor-Robinson D, Webster ADB. Mycoplasmas and ureaplasmas in patients with hypogammaglobulinemia and their role in arthritis: microbiological observations over twenty years. Ann Rheum Dis 1994;53:183-7.

Gaasterland, T. and Sensen, C., W. 1996. Fully automated genome analysis that reflects user needs and preferences - a detailed introduction to the MAGPIE system architecture. Biochimie, 78, 302-310.

Gamier, M., Danel, N., Bové, J. M. 1984. The greening organism is a gram negative bacterium. See Ref. 17, pp. 115-24.

Garnier, M. 1997. Non-cultivable phytopathogenic mycoplasmas: characterization, detection and perspectives for control. Wiener Klinische Wochenschrift 109:613-617.

Garnier, M., M. Clerc, and J.-M. Bove. 1981. Growth and division of spiroplasmas: morphology of Spiroplasma citri during growth in liquid medium. J. Bacteriol. 147:642-652.

Gaurivaud, P., Danet, J.L., Bové, J.M., and Laigret, F. (1998) Fructose metabolism and pathogenicity of Spiroplasma citri. Pages 197-198. In: Program and Abstracts, 12th International Organization of Mycoplasmology Conference. 22-28 July 1998.

Gerstenecker B, Jacobs E. Topographical mapping of the P1-adhesin of Mycoplasma pneumoniae with adherence-inhibiting monoclonal antibodies. J Gen Microbiol 1990;136:471-6.

Ghosh, S. K., Raychaudhuri, S. P., Varma, A., Nariani, T. K. 1971. Isolation and culture of mycoplasma associated with citrus greening disease. Curr. Sci. 40:299-300.

Gignac SM, Brauer S, Häne B, Quentmeier H, Drexler HG: Elimination of mycoplasma from infected leukemia cell lines. Leukemia 5: 162-165 (1991).

Gignac SM, Uphoff CC, MacLeod RAF, Steube K, Voges M, Drexler HG: Treatment of mycoplasma-contaminated continuous cell lines with Mycoplasma Removal Agent (MRA). Leukemia Res. 16: 815-822 (1992).

Girón JA, Lange M, Baseman JB. Adherence, fibronectin binding, and induction of cytoskeleton reorganization in cultured human cells by Mycoplasma penetrans. Infect Immun 1996;64:197-208.

Glew, M. D., Marenda, M., Rosengarten, R., Citti, C. (2002). Surface Diversity in Mycoplasma agalactiae Is Driven by Site-Specific DNA Inversions within the vpma Multigene Locus. J. Bacteriol. 184: 5987-5998.

Goffeau, A., Barrell, B. G., Bussey, H., Davis, R. W., Dujon, B. et al. (1996). Life with 6000 genes. Science, 274, 546-567.

Goidýnich, G. 1959. Manual di Patologia Vegetale. Edizioni Agricole, Bologna. 713 pp.

Golino, D.A. G.N. Oldfield and D.J. Gumpf: Phytopathology 77, 954-957 (1987)

Golino, D.A., G.N. Oldfield and D.J. Gumpf: Phytopathology 78, 285-288 (1988)

Goodwin, P. H., Xue, B. G., Kuske, C. R. and Sears, M. K. 1994. Amplification of plasmid DNA to detect plant pathogenic mycoplasmalike organisms. Ann. Appl. Biol. 124: 27-36.

Goody, J. B., M. T. Franklin and D. J. Hooper. 1965. The nematode parasites of plants catalogued under their hosts. Commonwealth Agricultural Bureaux, Farnham Royal Co., Bucks, U.K. 214 pp.

Goswami, B. K., Raychaudhuri, S. P., Nariani, T. K. 1971. Free amino acid content of the greening-affected and healthy plants of sweet orange (Citrus sinensis Osbeck). Curr. Sci. 17:469-70.

Goulet M, Dular R, Tully JG, Billowes G, Kasatiya S. Isolation of Mycoplasma pneumoniae from the human urogenital tract. J Clin Microbiol 1995;33:2823-2825.

Grau O, Slizewicz B, Tuppin P, Launay V, Bourgeois E, Sagot N, et al.1995. Association of Mycoplasma penetrans with human immunodeficiency virus infection. J Infect Dis;172:672-81.

Grau, C.R. 1990. Aphanomyces root rot. Pages 10-11 in Compendium of Alfalfa Diseases: Second Edition, D.L. Stuteville and D.C. Erwin, eds. APS Press, St. Paul, MN.

Grau, C.R. 1992. Registration of WAPH-1 alfalfa germplasm with resistance to Aphanomyces root rot. Crop Sci. 32: 287-288.

Griffiths, H. M., Gundersen, D. E., Sinclair, W. A., Lee, I-M., and R. E. Davis. 1994a. Mycoplasmalike organisms from milkweed, goldenrod, and spirea represent two new 16S rRNA sub-groups and three new strain subclusters related to peach X-disease. Canadian Journal of Plant Pathology 16: 255-266.

Griffiths, H.M., Sinclair, W.A., Smart, C.D., and Davis, R.E.. (1999) The phytoplasma associated with ash yellows and lilac witches'-broom: 'Candidatus Phytoplasma fraxini". Int. J. Syst. Bacteriol. (in press).

Gundersen, D. E., Lee, I-M., Rehner, S. A., Davis, R. E. and D. T. Kingsbury. 1994. Phylogeny of mycoplasmalike organisms (phytoplasmas): a basis for their classification. Journal of Bacteriology 176: 5244-5254.

Gunderson, D. E. and I. M. Lee. 1996. Ultrasensitive detection of phytoplasmas by nested-PCR assays using two universal primer pairs. Phytopath. Medit. 35:144-151.

Gunderson, D. E., I. M. Lee, D. A. Schaff, N. A. Harrison, C. J. Chang, R. E. Davis, D. T. Kingsbury. 1996. Genomic diversity and differentiation investigated by nucleic acid hybridization and restriction fragment length polymorphism analyses. Appl. Environ. Micro. 57 (12): 3565-3569.

Guo, Y. H., Cheng, Z. M., Walla, J. A. and Z. Zhang. 1998. Diagnosis of X-disease phytoplasma in stone fruits by a monoclonal antibody developed directly from a woody plant. Journal of Environmental Horticulture 16: 33-37.

Guo, Y.H., Z.M. Cheng, J.A. Walla, and Z. Zhang. 1998. Large-scale screening for X-disease phytoplasma infection in chokecherry. HortScience 33:293-295.

Hanor, H., D. Goodman, and G. S. Stent. 1969. RNA chain growth rates in Escherichia coli. J. Mol. Biol. 39:1-20.

Haque, S. Q. and S. Parasram. 1973. Empoasca stevensi, A New Vector of Bunchy Top Disease of Papaya. Plant Dis. Reptr. 57(5): 412-413.

Hartowicz, L. E. et al. 1971. Possible biocontrol of wild hemp. North Central Weed Control Conference, Proceedings 26:69.

He, F.-T., Lee, I.-M., Hammond, R.W., and Zhao, Y. 1998. Preliminary investigation of the role of protein kinase gene expression in the induction of phyllody symptoms on plants infected with phytoplasmas. Pages 184-185. In: Program and Abstracts, 12th International Organization of Mycoplasmology Conference. 22-28 July 1998.

Helmstetter, F. C. 1996. Timing of synthetic activities in the cell cycle, p. 1627-1639. In F. C. Neidhardt, R. Curtiss III, J. L. Ingraham, E. C. C. Lin, K. B. Low, B. Magasanik, W. S. Reznikoff, M. Riley, M. Schaechter, and H. E. Umbarger (ed.), Escherichia coli and Salmonella: cellular and molecular biology, 2nd ed. ASM Press, Washington, D.C.

Himmelreich, R., H. Hilbert, H. Plagens, E. Pirkl, B.-C. Li, and R. Herrmann. 1996. Complete sequence analysis of the genome of the bacterium Mycoplasma pneumoniae. Nucleic Acids Res. 24:4420-4449.

Hockey, J. F. 1927. Report of the Dominion field laboratory of plant pathology, Kentville, Nova Scotia. Canada Department of Agriculture: 28-36.

Hopert A, Uphoff CC, Wirth M, Hauser H, Drexler HG: Mycoplasma detection by PCR analysis. In Vitro Cell Dev Biol 29A: 819-821 (1993).

Hopert A, Uphoff CC, Wirth M, Hauser H, Drexler HG: Specificity and sensitivity of polymerase chain reaction (PCR) in comparison with other methods for the detection of mycoplasma contamination in cell lines. J Immunol Methods 164: 91-100 (1993).

Hu PC, Schaper U, Collier AM, Clyde WA, Horikawa M, Huang YS, et al. 1987. A Mycoplasma genitalium protein resembling the Mycoplasma pneumoniae attachment protein. Infect Immun 55:1126-31.

Humphery-Smith, I., C. Chastel, O. Grulet, and F. LeGoff. (1990) The infective cycle of Spiroplasma sabaudiense in vitro in Aedes albopictus (C636) cells and its potential for the biological control of Dengue virus. Zbl. Bakt. Suppl. 20:922-924.

Inamine JM, Loechel S, Gilbert AM, Barile MF, Hu PC. Nucleotide sequence of the MgPa (mgp) operon of Mycoplasma genitalium and comparison to the P1 (mpp) operon of Mycoplasma pneumoniae. Gene 1989;82:259-67.

International Committee on Systematic Bacteriology, Sub-committee on the Taxonomy of Mollicutes. 1997. Minutes of the Interim meetings, 12 and 18 August, 1996, Orlando, Florida, USA International Journal of Systematic Bacteriology 47: 911-914.

Jacob C., F. Nouzieres, S. Duret, J. Bové and J. Renaudin 1997. Isolation, characterization and complementation of a motility mutant of Spiroplasma citri. J. Bacteriol., 179 (15), 4802-4810.

Jacoli, G.G. 1981. Attempts to culture in vitro mycoplasma-like organisms from plants: a retrospective view. Phytopathol Z. 102:148-152.

Jagoueix, S., Scott, P., Danet, J.L., Bové, J.M., and Garnier, M. (1998) Interaction of mollicutes with their host plant: Differential expression of plant genes following jinfection by mollicutes. Pages 196-197. In: Program and Abstracts, 12th International Organization of Mycoplasmology Conference. 22-28 July 1998.

Jones, A. L., J. L. Norelli, and G. R. Ehret. 1991. Detection of streptomycin-resistant Pseudomonas syringae pv. papulans in Michigan apple orchards. Plant Dis. 75:529-531.

Jung, H.-Y., Sawayanagi, T., Kakizawa, T., Nishigawa, H., Oshima, K., Miyata, S., Kuboyama, T., Ugaki, M. and Namba, S. (2001). Phylogenetic classification of the phytoplasma; Identification of a new phytoplasma from diseased chestnut tree in Korea. Japanese Journal of Mycoplasmology 28: 19-21.

Kakizawa, S., Furuki, K., Komoto, A., Sawayanagi, T., Oshima, K., Miyata, S., Kuboyama T., Ugaki, M., and Namba, S. (2001a). Cloning of a phytoplasmal membrane protein, its overexpression in E. coli and specific antibody production of the protein. Japanese Journal of Mycoplasmology 28: 44-45.

Kammer, G. M., J. D. Pollack, and A. S. Klainer. 1970. Scanning-beam electron microscopy of Mycoplasma pneumoniae. J. Bacteriol. 104:499-502.

Kaneko, T., Sato, S., Kotani, H., Tanaka, A., Asamizu, E. et al. 1996. Sequence analysis of the genome of the unicellular cyanobacterium Synechocystis sp. strain PCC6803. II. Sequence determination of the entire genome and assignment of potential protein-coding regions. DNA Res., 3, 109-136.

Kangro HO, Chong SKF, Hardiman A, Heath RB, Walker-Smith JA. A prospective study of viral and mycoplasma infections in chronic inflammatory bowel disease. Gastroenterology 1990;98:549-53.

Katseni VL, Gilroy CB, Ryait BK, Ariyoshi K, Bieniasz PB, Weber JN, et al. Mycoplasma fermentans in individuals seropositive and seronegative for HIV-1. Lancet 1993;341:271-3.

Khurana, S.M.P., R.A. Singh, and D.M. Kaley. 1988. Mycoplasma-associated potato diseases and their control in India. Pages 285-316 in Mycoplasma Diseases of Crops: Basic and Applied Aspects, K. Maramorosch and S.P. Raychaudhuri, eds. Springer-Verlag, NY. 456 pp.

Kirchhoff H, Rosegarten R, Lotz W, Fischer M, Lopatta D. Flask-shaped mycoplasmas: properties and pathogenicity for man and animals. Isr J Med Sci 1984;10:848-53.

Kiritani K and Su H. 1999. Papaya ring spot, banana bunchy top and Citrus greening in the Aisa Pacific Region: occurrence and control strategy. Japanese Journal of Tropical Agriculture 33(1): 23-30. (09128)

Kirkpatrick B.C., D.C. Stenger, T.J. Morris & A.H. Purcell (1987). Cloning and detection of DNA from a nonculturable plant pathogenic mycoplasma-like organism. Science, 238, 197-200.

Kirkpatrick, B. C., Smart, C., Gardner, S. Gao, J-L., Ahrens, U., Maurer, R., Schneider, B., Lorenz, K-H., Seemüller, E., Harrison, N., Namba, S. and X. Daire. 1994. Phylogenetic relationships of plant pathogenic MLOs established by 16/23S rDNA spacer sequences. IOM Letters 3: 228-229.

Kison, H., Schneider, B. and E. Seemüller. 1994. Restriction fragment length polymorphism within the apple proliferation mycoplasmalike organism. Journal of Phytopathology 141: 395-401.

Kitajima, E.W., C.M. Chagas, and O.A. Crestani. 1986. Virus and mycoplasma-associated diseases of passionfruit in Brazil. Fitopatologia Brasileira 11: 409-432.

Klenk, H.-P., Clayton, R. A., Tomb, J.-F., White, O., Nelson, K. E. et al. (1997). The complete genome sequence of the hyperthermopholic, sulphate-reducing archaeon Archaeoglobus fulgidus. Nature, 390, 364-370.

Kollar A., E. Seemüller, F. Bonnet, C. Saillard & J.M. Bové (1990) - Isolation of the DNA of various plant pathogenic mycoplasma-like organisms from infected plants. Phytopathol., 80: 233-237.

Kollar, A. and E. Seemüller 1989. Base composition of the DNA of mycoplasmalike organisms associated with various plant diseases. Journal of Phytopathology 141: 395-401.

Koonin, E. V. and Ilyina, T. V. (1992). Geminiviruses replication proteins are related to prokaryotic plasmid rolling circle DNA replication initiator proteins. J. Gen. Virol. 73: 2763-2766.

Kosslak, R. M. and B. B. Bohlool 1983. Prevalence of Azospirillum spp. in the rhizosphere of tropical plants. Canadian Journal of Microbiology 29: 649-652.

Kranz, A.R. and B. Kirchheim: Genetic Resources in Arabidopsis, Arabid. Inf. Serv. (Frankfurt am Main) 24, (1987)

Krause DC, Taylor-Robinson D. Mycoplasmas which infect humans. In: Maniloff J, McElhaney RN, Finch LR, Baseman JB, editors. Mycoplasmas: molecular biology and pathogenesis. Washington (DC): American Society for Microbiology, 1992:417-44.

Krause DC. Mycoplasma pneumoniae cytadherence: unravelling the tie that binds. Mol Microbiol 1996;20:247-53.

Kuboyama, T., Huang, C.-C. Lu, X. Sawayanagi, T. Kanazawa, T. Kagami, T. Matsuda, I. Tsuchizaki, T. and Namba, S. (1998). A plasmid isolated from phytopathogenic Onion Yellows phytoplasma and its heterogeneity in the pathogenic phytoplasma mutant. Mol. Plant-Microbe Interact. 11: 1031-1037.

Kunst, F., Ogasawara, N., Moszer, I., Albertini, A. M., Alloni, G. et al. (1997). The complete genome sequence of the Gram-positive bacterium Bacillus subtilis. Nature, 390, 249-256.

Kuske, C. R. and B. C. Kirkpatrick. 1992. Phylogenetic relationships between the western aster yellows amycoplasmalike organism and other prokaryotes established by 16S rRNA gene sequence. International Journal of Systematic Bacteriology 42: 226-233.

Kuske, C. R. and Kirkpatrick, B. C. (1990). Identification and characterization of plasmids from the western aster yellows mycoplasmalike organism. J. Bacteriol. 172: 1628-1633.

Layh-Schmitt G, Hilbert H, Pirkl E. A spontaneous hemadsorption-negative mutant of Mycoplasma pneumoniae exhibits a truncated adhesin-related 30-kilodalton protein and lacks the cytadherence-accessory protein HMW1. J Bacteriol 1995;177:843-6.

Lee I.-M., D.E. Gundersen-Rindal and A. Bertaccini (1998) - Phytoplasmas: ecology and genomic diversity. Phytopathol., 88 (12), 1359-1366.

Lee, I. M., R. W. Hammond, R. E. Davis, D. E. Gunderson. 1993. Universal amplification and analysis of pathogen 16S rDNA for classification and identification of mycoplasmalike organisms. Phytopathology. 83: 834-842.

Lee, I.-M., A. Bertaccini, M. Vibio, and D.E. Gunderson. 1995. Detection of multiple phytoplasmas in perennial fruit trees with decline symptoms in Italy. Phytopathology 85: 728-735.

Lee, I.M., D.E. Gundersen-Rindal, R.E. Davis, and I.M. Bartoszyk. 1998. Revised classification scheme of phytoplasmas based on RFLP analyses of 16S rRNA and ribosomal protein gene sequences [Review]. International Journal of Systematic Bacteriology. 48:1153-1169.

Lee, I-M., Davis, R. E., Chen, T-A., Chiykowski, L. N., Fletcher, J., Hiruki, C. and D. A. Schaff. 1992. A genotype-based system for identification and classification of mycoplasmalike organisms (MLOs) in the aster yellows MLO strain cluster. Phytopathology 82: 977-986.

Lefol, C., J. Lherminier, E. Boudon-Padieu, R. Meignoz, J. Larrue, C. Louis, A.-C. Roche, and A. Caudwell. 1995. Presence of attachment sites accounting for recognition between the flavescence dorée MLO and its leafhopper vector. IOM Letters 3: 282-283.

Lichtarge, O., Bourne, H. R. and Cohen, F. E. 1996. An evolutionary trace method defines binding surfaces common to protein families. J. Mol. Biol., 257, 342-358.

Lim, P.O. and Sears, B.B. 1989. 16S rRNA sequence indicates that plant-pathogenic mycoplasmalike organisms are evolutionarily distinct from animal mycoplasmas. J. Bacteriol. 171: 5901-5906.

Lim, P.O.and Sears, B.B. 1992. Evolutionary relationships of a plant-pathogenic mycoplasmalike organism and Acholeplasma laidlawii deduced from two ribosomal protein gene sequences. J. Bacteriol. 174:2606-2611.

Lima, C. D., Klein, M. G. and Hendrickson, W. A.1997. Structure-based analysis of catalysis and substrate definition in the HIT protein family. Science, 278, 286-290.

Lin, C.P., and T.A. Chen. 1986. Comparison of Monoclonal antibodies and polyclonal antibodies in detection of the aster yellows mycoplasmalike organism. Phytopathology. 76:45-50.

Lo SC, Dawson MS, Newton III PB, Sonoda MA, Shih JW, Engler WF, et al. Association of the virus-like infectious agent originally reported in patients with AIDS with acute fatal disease in previously healthy non-AIDS patients. Amer J Trop Med Hyg 1989;41:364-76.

Loper, J. E., M. D. Henkels, R. G. Roberts, G. G. Grove, M. J. Willett, and T. J. Smith. 1991. Evaluation of streptomycin, oxytetracycline, and copper resistance of Erwinia amylovora isolated from pear orchards in Washington State. Plant Disease 75:287-290.

Luttrell LM, Kanj SS, Corey GR, Lins RE, Spinner RJ, Mallon WJ, et al. Mycoplasma hominis septic arthritis: two case reports and review. Clin Infect Dis 1994;19:1067-70.

Macpherson I, Russell W. Transformations in hamster cells mediated by mycoplasmas. Nature 1966;210:1343-5.

Malvick, D.K. 1998. Aphanomyces Race 2 is distributed widely in Wisconsin and the United States. Page 1 in The Haymaker: Summer 1998. W-L Research, Inc., Evansville, WI.

Malvick, D.K., C.R. Grau, and J.A. Percich. 1998. Characterization of Aphanomyces euteiches strains based on pathogenicity tests and random amplified polymorphic DNA analyses. Mycol. Res. 102: 465-475.

Marcone, C., H. Neimark, and E. Seemuller. 1999. Genome sizes of phytoplasmas composing major phylogenetic groups and subgroups. Phytopathol. 89:105-110.

Markham, P.G. 1982. The "yellows" plant diseases: plant hosts and their interaction with the pathogens., pp. 82-100. In M.J. Daniels and P.G. Markham (ed.), Plant and Insect mycoplasma techniques. Croom Helm, London.

Matsuoka, K. and M.G. Carvalho. 1987. Mycoplasma-like organisms associated with kaki decline in Brazil. Plant Pathology 36: 417-419.

McClean, A. P. D. 1970. Greening disease of sweet orange: its transmission in propagative parts and distribution in partially diseased trees. Phytophylactica 2:263-68.

McCoy R.E., A. Caudwell, C.J. Chang, T.A. Chen, L.N. Chiykowski, M.T. Cousin, et al. (1989) - Plant diseases associated with mycoplasma-like organisms. in R.F. Whitcomb, J.G. Tully (eds). The mycoplasmas, vol 5, Spiroplasmas, acholeplasmas, and mycoplasmas of plants and arthropods. Academic Press,New York, pp. 545-640.

McGarrity GJ, Kotani H, Butler GH. Mycoplasmas and tissue culture cells. In: Maniloff J, McElhaney RN, Finch LR, Baseman JB, editors. Mycoplasmas: molecular biology and pathogenesis. Washington (DC): American Society for Microbiology, 1992:445-454.

McManus, P. S., and A. L. Jones. 1994. Epidemiology and genetic analysis of streptomycin-resistant Erwinia amylovora from Michigan and evaluation of oxytetracycline for control. Phytopathology 84:627-633.

McPartland, J. M. 1991. Common names for diseases of Cannabis sativa L. Plant Disease 75: 226-227.

McPartland, J. M. 1994. Microbiological contaminants of marijuana. J International Hemp Association 1(1): 41-44.

McPartland, J. M. 1995c. Cannabis pathogens X: Phoma, Ascochyta and Didymella species. Mycologia 86:870-878.

McPartland, J. M. and D. F. Schoeneweiss 1984. Hyphal morphology of Botryosphaeria dothidea in vessels of unstressed and drought-stressed Betula alba. Phytopathology 74: 358-362.

McPartland, J. M. and S. Hughes 1994. Cannabis pathogens VII: a new species, Schiffnerula cannabis. Mycologia 86: 867-869.

Minion, F. C., K. J. Jarvill-Taylor, D. E. Billings, and E. Tigges. 1993. Membrane-associated nuclease activities in mycoplasmas. J. Bacteriol. 175:7842-7847.

Minsavage, G. V., B. I. Canteros, and R. E. Stall. 1990. Plasmid-mediated resistance to streptomycin in Xanthomonas campestris pv. vesicatoria. Phytopathology 80:719-723.

Mitchell, J.E. and C.Y. Yang. 1966. Factors affecting growth and development of Aphanomyces euteiches. Phytopathology 56: 917-922.

Miyata, M., and T. Fukumura. 1997. Asymmetrical progression of replication forks just after initiation on Mycoplasma capricolum chromosome revealed by two-dimensional gel electrophoresis. Gene 193:39-47.

Miyata, M., Yamamoto, H., Shimizu, T., Uenoyama, A., Citti, C., Rosengarten, R. 2000. Gliding mutants of Mycoplasma mobile: relationships between motility and cell morphology, cell adhesion and microcolony formation. Microbiology 146: 1311-1320.

Miyata, S., Furuki, K., Sawayanagi, T., Oshima, K., Kuboyama, T., Tsuchizaki, T., Ugaki, M. and Namba, S. 2002. Gene arrangement and sequence of str operon of phytoplasma resemble those of Bacillus more than those of Mycoplasma. Journal of General Plant Pathology 68: 62-67.

Mokhbat JE, Person PK, Sabath LD, Robertson JA. Peritonitis due to Mycoplasma hominis in a renal transplant patient. J Infect Dis 1982;146:713.

Moller, W. J., M. N. Schroth, and S. V. Thomson. 1981. The scenario of fire blight and streptomycin resistance. Plant Disease 65:563-568.

Morowitz, H. J., and J. Maniloff. 1966. Analysis of the life cycle of Mycoplasma gallisepticum. J. Bacteriol. 91:1638-1644.

Murray HW, Masur H, Senterfit LB, Roberts RB. The protean manifestations of Mycoplasma pneumoniae infection in adults. Am J Med 1975;58:229-42.

Naidu, R., Govindu, H. C. 1981. Strains and strain interactions of the citrus greening pathogen in India. In Mycoplasma Diseases of Trees and Shrubs, ed. K. Maramarosch, S. P. Raychaudhuri, pp. 299-313. New York: Academic.

Nakashima, K. and Hayashi, T. (1997). Sequence analysis of extrachromosomal DNA of sugarcane white leaf phytoplasma. Ann. Phytopathol. Soc. Jpn. 63: 21-25.

Namba, S., Kato, S., Iwanami, S., Oyaizu, H., Shiozawa, H. and Tsuchizaki, T. (1993a). Detection and differentiation of plant-pathogenic mycoplasma-like organisms using polymerase chain reaction. Phytopathology 83: 786-791.

Namba, S., Oyaizu, H., Kato, S., Iwanami, S. and Tsuchizaki, T. (1993b). Phylogenetic diversity of phytopathogenic mycoplasmalike organisms. Int. J. Syst. Bacteriol. 43: 461-467.

Neale, G. A. M., A. Mitchell, and L. R. Finch. 1983. Pathways of pyrimidine deoxyribonucleotide biosynthesis in Mycoplasma mycoides subsp. mycoides. J. Bacteriol. 154:17-22.

Neimark, H. 2001. Haemotrophic Mycoplasmas (Containing the former rickettsia genera Haemobartonella and Eperythrozoon). In: The encyclopedia of arthropod-transmitted infections. Edited by M.W. Service. Wallingford, U.K. CABI Publishing.

Neimark, H., Barnaud, A., Gounon, P., Michel, J.C., Contamin, H. 2002. The putative haemobartonella that influences Plasmodium falciparum parasitaemia in squirrel monkeys is a haemotrophic mycoplasma. Microbes Infect. 4:693-698.

Neimark, H., K.-E. Johansson, Y. Rikihisa, and J.G. Tully. 2001. Proposal to transfer members of the genera Haemobartonella and Eperythrozoon to the genus Mycoplasma with descriptions of Candidatus Mycoplasma haemofelis, Candidatus Mycoplasma haemomuris, Candidatus Mycoplasma haemosuis and Candidatus Mycoplasma wenyonii. Internat. J. Syst. Evolut. Microbiol. 51:891-899.

Niblett, C.L., A.C. Monllor, R. Provvidenti, P. E. Still, J. Escudero, J. Bird, and K. Gough. 1991. A disease of passionfruit (Passiflora edulis) in Puerto Rico, possibly caused by an isolate of water-melon mosaic virus 2. Phytopathology 81: 698.

Nicolson G, Nicolson NL. Diagnosis and treatment of mycoplasmal infections in Gulf War illness-CFIDS patients. Intl J Occup Med Immunol Toxicol 1996;5:69-78.

Nishigawa, H., Miyata, S. Oshima, K., Sawayanagi, T., Komoto, A., Kuboyama, T., Matsuda, I., Tsuchizaki, T. and Namba S. (2001a). In planta expression of a protein encoded by the extrachromosomal DNA of a phytoplasma and related to geminivirus replication proteins. Microbiology 147: 507-513.

Nishigawa, H., Oshima, K., Kakizawa, S., Jung, H.-Y., Kuboyama, T., Miyata, S., Ugaki, M. and Namba, S. (2002). A plasmid from a non-insect-transmissible line of a phytoplasma lacks two open reading frames that exist in the plasmid from the wild-type line. Gene 298: 195-201.

Nishimura, N., Nakajima, S., Sawayanagi, T. and Namba, S. (1998). Transmission of Cryptotaenia japonica witches' broom phytoplasma and oinion yellows phytoplasma by Hishimonus sellatus Uhler. Ann. Phytopath. Soc. Japan 64: 474-477.

Oliveira, C.R.B., V.L.A. Marinho, F.S. Astolfi, M. Azevedo, C.M. Chagas, and E.W. Kitajima. 1994. Purification, serology and some properties of the purple granadilla (Passiflora edulis) mosaic virus. Fitopatologia Brasileira 19: 455-462.

Oshima, K., Kakizawa, S., Nishigawa, H., Kuboyama, T., Miyata, S., Ugaki, M. and Namba, S.2001. A plasmid of phytoplasma encodes a unique replication protein having both plasmid- and virus-like domains: clue to viral ancestry or result of virus/plasmid recombination? Virology 285: 270-277.

Oshima, K., Shiomi, T., Kuboyama, T., Sawayanagi, T., Nishigawa, H., Kakizawa, S., Miyata, S., Ugaki, M. and Namba, S. 2001. Isolation and characterization of derivative lines of the onion yellows phytoplasma that do not cause stunting or phloem hyperplasia. Phytopathology, 9: 1024-1029.

Palmer, E. L., and Jones, A. L. 1997. Tetracycline-resistance determinants present in Michigan apple orchards. Phytopathology 87:S74.

Pandotra, V. R. and K. S. M. Sastry 1967. Wilt: a new disease of hemp in India. Indian Journal of Agricultural Science 37: 520.

Panopoulos, N. and Peet, R. C. (1985). The molecular genetics of plant pathogenic bacteria and their plasmids. Annu Rev Phytopathol 23: 381-419.

Pares, R.D., A.B. Martin, and R.D. Fitzell. 1985. Virus-induced tip necrosis of passionfruit (Passiflora edulis Sims). Australasian Plant Pathology 14: 76-78.

Paton GR, Jacobs JP, Perkins FT. 1966. Chromosome changes in human diploid-cell cultures infected with Mycoplasma. Nature 207:43-5.

Pfender, W.F., P.A. Delwiche, C.R. Grau, and D.J. Hagedorn. 1984. A medium to enhance recovery of Aphanomyces from infected plant tissue. Plant Dis. 68: 845-847.

Phatak, H. C. et al. 1975. Mycoplasma-like bodies associated with Cannabis phyllody. Phytopath. Z. 83: 281-284.

Ploaie, P. G. 1981. Mycoplasma-like organisms and plant diseases in Europe. Pages 61-104 In Plant Diseases and Vectors: Ecology and Epidemiology. Maramorosch, K., and Harris, K. F., eds. Academic Press, New York.

Provvidenti, R. 1996. Disease caused by a phytoplasma: Aster yellows. Page 37 in Compendium of Cucurbit Diseases, T.A. Zitter, D.L. Hopkins, and C.E. Thomas, eds. APS Press, St. Paul, MN 87 pp.

Purohit A G. 1993. Papaya nutrition. Advances in Horticulture: Fruit Crops - Volume 2: New Delhi, India: Malhotra Publishing. pp. 907-913. (09637/F01.C42)

Pyle, L. E., and L. R. Finch. 1988. A physical map of the genome of Mycoplasma mycoides subspecies mycoides Y with some functional loci. Nucleic Acids Res. 16:6027-6039.

Pyle, L. E., and L. R. Finch. 1988. Preparation and FIGE separation of infrequent restriction fragments from Mycoplasma mycoides DNA. Nucleic Acids Res. 16:2263-2268.

Quardokus, E., N. Din, and Y. V. Brun. 1996. Cell cycle regulation and cell type-specific localization of the FtsZ division initiation protein in Caulobacter. Proc. Natl. Acad. Sci. USA 93:6314-6319.

Ram M. 1993. Improvement of papaya. Advances in Horticulture; Fruit Crops - Volume 1: New Delhi, India: Malhotra Publishing House. pp. 383-397. (09279/1100.C43)

Rascoe, J., U. Melcher, and J. Fletcher. (1998) Arbitrarily primed PCR of cDNA for determination of differential gene presence and expression among lines of Spiroplasma citri. Phytopathology 88:S75.

Rawadi G, Roman-Roman S. 1996. Mycoplasma membrane lipoproteins induce proinflammatory cytokines by a mechanism distinct from that of lipopolysaccharide. Infect Immun ;64:637-43.

Raychaudhuri, S. P., Nariani, T. K., Ghosh, S. K., Viswanath, S. M., Kumar, D. 1974. Recent studies on citrus greening in India. See Ref. 27, pp. 53-57.

Razin S., D. Yogev & Y. Naot 1998. Molecular biology and pathogenicity of mycoplasmas. Microbiol. Molec. Biol. Rev., 62 (4), 1094-1156.

Reddy SP, Rasmussen WG, Baseman JB. 1995. Correlations between Mycoplasma pneumoniae sensitivity to cyclosporin A and cyclophilin-mediated regulation of mycoplasma cytadherence. Microb Pathog 20:155-69.

Reddy SP, Rasmussen WG, Baseman JB. 1995. Molecular cloning and characterization of an adherence-related operon of Mycoplasma genitalium. J Bacteriol 177:5943-5951.

Renaudin J., A. Marais, E. Verdin, S. Duret, X. Foissac, F. Laigret and J.M. Bové 1995. Integrative and free Spiroplasma citri oriC plasmids. Expression of the Spiroplasma phoeniceum spiralin in Spiroplasma citri. J. Bacteriol., 177 (10), 2870-2877.

Rodwell, A. W. 1983. Defined and partly defined media, p. 163-172. In S. Razin, and J. S. Tully (ed.), Methods in mycoplasmology. Academic Press, New York, N.Y.

Root-Bernstein RS, Hobbs SH. Homologies between mycoplasma adhesion peptide, CD4 and class II MHC proteins: a possible mechanism for HIV-mycoplasma synergism in AIDS. Res Immunol 1991;142:519-23.

Rosengarten, R., and H. Kirchhoff. 1989. Growth morphology of Mycoplasma mobile 163K on solid surfaces: reproduction, aggregation, and microcolony formation. Curr. Microbiol. 18:15-22.

Rottem S and Barile MF.1993. Beware of Mycoplasmas. TIBTECH , Vol. 11:143-151.

Sato, M., Namba, S., Katsuhara, M., Kawakita, H., Mitsuhashi, W. and Kawabe, S. 1996. PCR (polymerase chain reaction) detection of phytoplasma in phloem sap collected by laser stylectomy. Ann. Phytopath. Soc. Japan 62 : 177-180.

Sawayanagi, T., Horikoshi, N., Kanehira, T., Shinohara, M., Bertaccini, A., Cousin, M.T., Hiruki, C. and Namba, S. 1999. 'Candidatus phytoplasma japonicum', a new phytoplasma taxon associated with Japanese Hydrangea phyllody. Int. J. Syst. Bacteriol. 49: 1275-85.

Schaeverbeke T, Gilroy CB, Bébéar C, Dehais J, Taylor-Robinson D.1996. Mycoplasma fermentans, but not M. penetrans, detected by PCR assays in synovium from patients with rheumatoid arthritis and other rheumatic disorders. J Clin Pathol. 41:311-4.

Schaff, D., I. M. Lee, R. E. Davis. 1992. Sensitive detection and identification of mycoplasma-like organisms in plants by polymerase chain reactions. Biochem. Biophys. Res. Comm. 186: 1503-1509.

Scheck, H. J., J. W. Pscheidt, and L. W. Moore. 1996. Copper and streptomycin resistance in strains of Pseudomonas syringae from Pacific Northwest Nurseries. Plant Dis. 80:1034-1039.

Schneider, B. & Seemüller, E. (1994). Presence of two sets of ribosomal genes in phytopathogenic mollicutes.Appl Environ Microbiol 60, 3409-3412.

Schneider, B., K.S. Gibb, and E. Seemuller. 1997. Sequence and RFLP analysis of the elongation factor Tu gene used in differentiation and classification of phytoplasmas. Microbiology. 143: 3381-3389.

Sears, B. B., Lim, P. O., Holland, N., Kirkpatrick, B. C. and K. L. Klomparens. 1989. Isolation and characterization of of DNA from a mycoplasma-like organism. Molecular Plant-Microbe Interactions 2: 175-180.

Seemuller, E., B. Schneider, R. Maurer & 8 other authors. 1994. Phylogenetic classification of phytopathogenic mollicutes by sequence analysis of 16S ribosomal DNA. Int J. Syst. Bacteriol. 44: 440-446.

Seemuller, E., K.-H. Lorenz and U. Lauer. 1998. Fruit tree phytoplasmas - Pear decline resistance in Pyrus communis rootstocks and progenies of wild and ornamental Pyrus taxa. In Hadidi, A., ed. Proc. 17th Int. Symp. on Fruit Tree Virus Dis. Acta Hort. 472:681-687.

Seemüller, E., Marcone, C., Lauer, U., Ragozzino, A., and M. Göschl. 1998. Current status of molecular classification of the phytoplasmas. Journal of Plant Pathology 80: 3-26.

Seemüller, E., Schneider, B., Maurer, R., Ahrens, U., Daire, X., Kison, H., Lorenz, K.H., Firrao, G., Avinent, L., Sears, B.B. and et, al. 1994. Phylogenetic classification of phytopathogenic mollicutes by sequence analysis of 16S ribosomal DNA. Int. J. Syst. Bacteriol. 44: 440-446.

Seto, S., Layh-Schmitt, G., Kenri, T., Miyata, M. 2001. Visualization of the Attachment Organelle and Cytadherence Proteins of Mycoplasma pneumoniae by Immunofluorescence Microscopy. J. Bacteriol. 183: 1621-1630.

Shaffer, W. H., and R. N. Goodman. 1985. Appearance of streptomycin-resistant Erwinia amylovora in Missouri apple orchards. Phytopathology 75:1281.

Sharma, G. K. and S. K. Mann 1984. Variation in vegetative growth and trichomes in Cannabis sativa L. (marihuana) in response to environmental pollution. J. Tennessee Academy of Science 59: 38-40.

Shiomi, T., Tanaka, M. and Wakiya, H. and, Zenbayashi, R. (1996). Occurrence of welsh onion yellows. Ann. Phytopathol. Soc. Jpn. 62: 258-260.

Siddique, A.B.M., J.N. Guthrie, K.B. Walsh, D.T. White, and P.T. Scott. 1998. Histopathology and within-plant distribution of the phytoplasma associated with Australian papaya dieback. Plant dis. 82(10):1112-1120.

Sinclair, W.A. 1995. Epidemiology of a slow-decline phytoplasmal disease: ash yellows on old field sites in New York State. Phytopathology 85: 123-128.

Sinclair, W.A., H.M. Griffiths, and I.-M. Lee. 1994. Mycoplasmalike organisms as causes of slow growth and decline of trees and shrubs. J. Aboriculture 20: 176-189.

Sitnik, V. P. 1981. Inheritance of charcters controlled by the pleiotropic effect of the gene for yellow stem in hemp. Selektsiya i Semenovodstvo No. 47: 46-49.

Smith, D. R., Doucette-Stamm, L. A., Deloughery, C., Lee, H., Dubois, J. et al. 1997. Complete genome sequence of Methanobacterium thermoautotrophicum delta H: functional analysis and comparative genomics. J. Bacteriol., 179, 7135-7155.

Stall, R. E., and P. L. Thayer. 1962. Streptomycin resistance of the bacterial spot pathogen and control with streptomycin. Plant Dis. Rep. 46:389-392.

Stewart, P. S., and R. D'Ari. 1992. Genetic and morphological characterization of an Escherichia coli chromosome segregation mutant. J. Bacteriol. 174:4513-4516.

Story, G. E., and Halliwell, R. S. 1969. Association of a mycoplasma-like organism with bunchy top disease of papaya. Phytopathology 59:1336-1337.

Su CJ, Chavoya A, Baseman JB. Regions of Mycoplasma pneumoniae cytadhesin P1 structural gene exist as multiple copies. Infect Immun 1988;56:3157-61.

Su CJ, Chavoya A, Dallo SF, Baseman JB. Sequence divergency of the cytadhesin gene of Mycoplasma pneumoniae. Infect Immun 1990;58:2669-74.

Sundin, G. W., and C. L. Bender. 1996. Dissemination of the strA-strB streptomycin resistance genes among commensal and pathogenic bacteria from humans, animals, and plants. Mol. Ecol. 5:133-143.

Synnott, K. [Ed.] 1941. Plant diseases. Notes contributed by the biological branch. Agricultural Gazette of New South Wales 52(7): 369-371, 384; 52(8): 435-438.

Taylor-Robinson D, Gilroy CB, Horowitz S, Horowitz J. Mycoplasma genitalium in the joints of two patients with arthritis. Eur J Clin Microbiol Infect Dis 1994;13:1066-9.

Taylor-Robinson D. Mycoplasmas in rheumatoid arthritis and other human arthritides. J Clin Pathol 1996;49 781-2.

Taylor-Robinson D. The history and role of Mycoplasma genitalium in sexually transmitted diseases. Genitourin Med 1995;71:1-8.

Teranaka, M., and K. Otsuka. 1973. Culture of the mycoplasma-like organisms associated with several witches' broom and aster yellows plants on artificial media. Bull. Coll. Agric. Utsunomiya Univ. 8:11-25.

Termorshuizen, A. J. 1991. Literatuuronderzoek over ziekten bij nieuwe potenti'le gewassen. IPO-DLO Rapport No. 91-08, Instituut voor Planteziektenkundig Onderzoek, Wageningen, The Netherlands. 18 pp.

Thomson, S. V., S. C. Gouk, J. L. Vanneste, C. N. Hale, and R. Clark. 1993. The presence of streptomycin resistant strains of Erwinia amylovora in New Zealand. Acta. Hortic. 338:223-225.

Tomb, J.-F., White, O., Kerlavage, A. r., Clayton, R. A., Sutton, G. G. et al. (1997). The complete genome sequence of the gastric pathogen Helicobacter pylori. Nature, 388, 539-547.

Tryon VV, Baseman JB. Pathogenic determinants and mechanisms. In: Maniloff J, McElhaney RN, Finch LR, Baseman JB, editors. Mycoplasmas: molecular biology and pathogenesis. Washington (DC): American Society for Microbiology, 1992:457-71.

Tsai S, Wear DJ, Shih JW-K, Lo SC. Mycoplasmas and oncogenesis:persistent infection and multistage malignant transformation. Proc Natl Acad Sci USA 1995;92:10197-201.

Tully JG, Rose DL, Baseman JB, Dallo SF, Lazzell AL, Davis CP. Mycoplasma pneumoniae and Mycoplasma genitalium mixture in synovial fluid isolate. J Clin Microbiol 1995;33:1851-5.

Uehara, T., Tanaka, M., Shiomi, T., Namba, S., Tsuchizaki, T. and Matsuda, I.. 1999. Histopathological studies on two symptom types of phytoplasma associated with lettuce yellows. Ann. Phytopath. Soc. Japan 65: 465-469.

Uphoff CC, Brauer S, Grunicke D, Gignac SM, MacLeod RAF, Quentmeier H, Steube K, Tümmler M, Voges M, Wagner B, Drexler HG.1992. Sensitivity and specificity of five different mycoplasma detection assays. Leukemia 6: 335-341.

Verma, A. K., Singh, B. P. 1977. Metabolic changes induced by greening, 272. tristeza and complex form in Kagzi lime, Citrus aurantifolia. Indian J. Exp. Biol. 15:811-14.

Vivian, A., Murillo, J. and Jackson, R. W. 2001. The roles of plasmids in phytopathogenic bacteria: mobile arsenals? Microbiology 147: 763-780.

Wang RY-H, Shih J W-K, Weiss SH, Grandinetti T, Pierce PF, Lange M, et al.1993. Mycoplasma penetrans infection in male homosexuals with AIDS: high seroprevalence and association with Kaposi's Sarcoma. Clin Infect Dis;17:724-9.

Wang, X., and J. Lutkenhaus. 1996. Characterization of the ftsZ gene from Mycoplasma pulmonis, an organism lacking a cell wall. J. Bacteriol. 178:2314-2319.

Warbrick, E. 1997. Two's company, three's a crowd: the yeast two hybrid system for mapping molecular interactions. Structure, 5, 13-17.

Wayadande, A.C. and J. Fletcher. 1995. Transmission of Spiroplasma citri lines and their ability to cross gut and salivary gland barriers within the leafhopper vector Circulifer tenellus. Phytopathology 85:1256-1259

Webb, R. R., and Davis, M. J. 1987. Unreliability of latex-flow test for diagnosis of bunchy-top of papaya caused by a mycoplasmalike organism. Plant Dis. 71:192.

Weisburg, W. G., J. G. Tully, D. L. Rose, J. P. Petzel, H. Oyaizu, D. Yang, L. Mandelco, J. Sechrest, T. G. Lawrence, J. Van Etten, J. Maniloff, and C. R. Woese. 1989. A phylogenetic analysis of the mycoplasmas: basis for their classification. J. Bacteriol. 171:6455-6467.

White, D. T., S. J. Billington, K. B. Walsh, and P. T. Scott. 1997. DNA sequence analysis supports the association of phytoplasmas with papaya (Carica papaya) dieback, yellow crinkle and mosaic. Australas. Plant Pathol. 26:28-36.

Wiersma, D.W., D.J. Undersander, J.G. Lauer, and C.R. Grau. 1997. Lack of alfalfa yield progress in the Midwest. 1997 Central Alfalfa Improvement Conference Proceedings.

Williams MH, Brostoff J, Roitt IM. Possible role of Mycoplasma fermentans in pathogenesis of rheumatoid arthritis. Lancet 1970;ii:270-80.

Ye F., F. Laigret, J.C. Whitley, C. Citti, L.R. Finch, P. Carle, J. Renaudin & J.M. Bové (1992) - A physical and genetic map of the Spiroplasma citri genome. Nucleic Acids Research, 20 (7), 1559-1565.

Ye, F., J. Renaudin, J.-M. Bove, and F. Laigret. 1994. Cloning and sequencing of the replication origin (oriC) of the Spiroplasma citri chromosome and construction of autonomously replicating artificial plasmids. Curr. Microbiol. 29:23-29.

Ye, F., Laigret, F., Carle, P. and Bové, J. M. 1995. Chromosomal heterogeneity among various strains of Spiroplasma citri.Int J Syst Bacteriol 45, 729-734.

Ye, F., U. Melcher, and J. Fletcher. 1997. Molecular characterization of a gene encoding a membrane protein of Spiroplasma citri. Gene 189:95-100.

Ye, F., Wayandande, A.C., Melcher, U., and Fletcher, J. 1998. Spiroplasma citri proteins adhering to organs of the leafhopper vector Circulifer tenellus. Pages 186-187. In: Program and Abstracts, 12th International Organization of Mycoplasmology Conference. 22-28 July 1998.

Yechouron A, Lefebre J, Robson HG, Rose DL, Tully JG. Fatal septicemia due to Mycoplasma arginini: a new human zoonosis. Clin Infect Dis 1992;15:434-8.

Yu Y.L., K.-W. Yeh & C.-P. Lin 1988. An antigenic protein gene of a phytoplasma associated with sweet potato witches'broom. Microbiol., 144, 1257-1262.

Yu, J., A.C. Wayadande, and J. Fletcher. 1997. Surface proteins of Spiroplasma citri involved in its adherence to vector insect cells. Phytopathology 87: S107.

Yu, Y.L., K.W. Yeh and C.P. Lin. 1998. An antigenic protein gene of a phytoplasma associated with sweet potato witches' broom. Microbiology 144:1257-1262.

Zaim M, Samad A. 1995. Association of phytoplasmas with a witches-broom diseaseof Withania somnifera (L.) Dunal in India. Plant Science 109. 9-225.

Zhang, X.G., Z.W Chen, Z.A. Chen, S.M. Sun, F.G. Zhang, and X.D. Tiang. 1991. Study on the infection of MLO of jujube witches' broom disease to periwinkle (Vinca rosea). International Journal of Tropical Plant Disease 9: 251-256.

Zhang, Y., J. K. Uyemoto, B. C. Kirkpatrick. 1998. A small-scale procedure for extracting nucleic acids from woody plants infected with various phytopathogens for PCR assay J. Virol. Methods. 71: 45-50

Zitter, T.A. 1991 Diseases caused by mycoplasmalike organisms: Aster yellows Page 43 in Compendium of Tomato Diseases, J.B. Jones. J.P Jones, R.E Stall, and T.A. Zitter, eds. APS Press, St. Paul, MN 73 pp

Zreik, L . Carle, P Bové, J.M., and Garnier, M (1995) Characterization of the mycoplasmalike organism associated with witches'-broom disease of lime and proposition of a Candidatus taxon for the organism, 'Candidatus Phytoplasma aurantifolia'. Int. J. Syst. Bacteriol 45 449-453.

INDEX